Experimental Researches

*On the Light and Luminous Matter of the
Glow-Worm, the Luminosity of the Sea, the
Phenomena of the Chameleon, the Ascent
of the Spider into the Atmosphere, and the
Torpidity of the Tortoise*

JOHN MURRAY

CAMBRIDGE
UNIVERSITY PRESS

CAMBRIDGE
UNIVERSITY PRESS

University Printing House, Cambridge, CB2 8BS, United Kingdom

Cambridge University Press is part of the University of Cambridge.
It furthers the University's mission by disseminating knowledge in the pursuit of
education, learning and research at the highest international levels of excellence.

www.cambridge.org
Information on this title: www.cambridge.org/9781108084031

© in this compilation Cambridge University Press 2017

This edition first published 1826
This digitally printed version 2017

ISBN 978-1-108-08403-1 Paperback

CAMBRIDGE LIBRARY COLLECTION

Books of enduring scholarly value

Zoology

Until the nineteenth century, the investigation of natural phenomena, plants and animals was considered either the preserve of elite scholars or a pastime for the leisured upper classes. As increasing academic rigour and systematisation was brought to the study of 'natural history', its subdisciplines were adopted into university curricula, and learned societies (such as the London Zoological Society, founded in 1826) were established to support research in these areas. These developments are reflected in the books reissued in this series, which describe the anatomy and characteristics of animals ranging from invertebrates to polar bears, fish to birds, in habitats from Arctic North America to the tropical forests of Malaysia. By the middle of the nineteenth century, this work and developments in research on fossils had resulted in the formulation of the theory of evolution.

Experimental Researches

John Murray (1785–1851), a writer and lecturer on many different scientific topics, published this collection of essays on what might be called the physics of biology in 1826. The first essay, on the luminosity of glow-worms, begins with an extensive discussion of the beauty and effects of light, and the various ways of creating it, before considering the various theories of light and optics current at the time. Supplied with specimens from Sweeny Hall in Shropshire, where they flourished, he performed various experiments on the 'luminous spherulae' which were the source of the glow-worm's light, trying to establish their chemical composition, and the time they would remain glowing in different media and temperatures. The same attention to detail and ingenious analysis are shown in the other studies, on the luminosity of the sea, the strength and lightness of spider webs, the chameleon's colour changes, and 'the torpidity of the tortoise'.

Cambridge University Press has long been a pioneer in the reissuing of out-of-print titles from its own backlist, producing digital reprints of books that are still sought after by scholars and students but could not be reprinted economically using traditional technology. The Cambridge Library Collection extends this activity to a wider range of books which are still of importance to researchers and professionals, either for the source material they contain, or as landmarks in the history of their academic discipline.

Drawing from the world-renowned collections in the Cambridge University Library and other partner libraries, and guided by the advice of experts in each subject area, Cambridge University Press is using state-of-the-art scanning machines in its own Printing House to capture the content of each book selected for inclusion. The files are processed to give a consistently clear, crisp image, and the books finished to the high quality standard for which the Press is recognised around the world. The latest print-on-demand technology ensures that the books will remain available indefinitely, and that orders for single or multiple copies can quickly be supplied.

The Cambridge Library Collection brings back to life books of enduring scholarly value (including out-of-copyright works originally issued by other publishers) across a wide range of disciplines in the humanities and social sciences and in science and technology.

EXPERIMENTAL RESEARCHES

ON

THE LIGHT AND LUMINOUS MATTER OF THE GLOW-WORM,

THE LUMINOSITY OF THE SEA,

THE PHENOMENA OF THE CHAMELEON,

THE

ASCENT OF THE SPIDER INTO THE ATMOSPHERE,

AND

THE TORPIDITY OF THE TORTOISE, &c.

By JOHN MURRAY, F. S. A., F. L. S.,

FELLOW OF THE HORTICULTURAL AND GEOLOGICAL SOCIETIES, MEMBER OF THE WERNERIAN AND METEOROLOGICAL SOCIETIES, &c. &c.

PRINTED FOR W. R. M'PHUN, GLASGOW;

DANIEL LIZARS, EDINBURGH;

BASIL STEUART, LONDON.

1826.

JAMES CURLL, PRINTER, GLASGOW.

TO

CHARLES MORLAND, Esq.

THESE PAGES

ARE RESPECTFULLY INSCRIBED.

ADVERTISEMENT.

A FEW of the subjects which the following pages embrace, have either already appeared, or been partially discussed. They are now introduced in a more condensed form and order, and exhibited at one view. New observations or illustrations have been added, and the phenomena altogether embrace some of the most curious and interesting inquiries to be found within the sublime and beautiful range of Natural History.

These are not mere speculative dogmas, but inferences resulting from inductive inquiries into the Physiology on which they treat, and they possess, at least, the charm of experimental research. Should they induce others to enter the field of investigation, and pursue the very singular phenomena thus partially unfolded, the object and design of the Author will be fully realized.

CONTENTS.

ON THE

LIGHT AND LUMINOUS MATTER

OF THE

"LAMPYRIS NOCTILUCA,"

OR

GLOW-WORM.

ON THE

LIGHT AND LUMINOUS MATTER

OF THE

"LAMPYRIS NOCTILUCA,"

OR

GLOW-WORM.

————

LIGHT kindles up the external world, and gives it a tongue to speak. It invests objects with a living character. Its rainbow-mantle clothes vegetation with all its beauty and enchantment, and the gems of the mineral kingdom reflect its prism's ray. In the absence of this power the beings of the vegetable kingdom languish, turn pale, and die, and even the province of zoology would sustain a shock which would shake it to its centre—would unnerve it.

On the question of the nature of light, two distinct and contrary opinions have prevailed, and are sustained by their respective adherents,

B

—that of Newton, who considered light as material, and consisting of particles, exceedingly minute, cast off from the luminous surface—while the other view of it, as held by Descartes and Euler, and maintained ingeniously enough by Drs. Young and Higgins, regarding light as a mere quality, infer it to be the result of the vibrations of a subtle medium pervading space, excited by a ray of the luminous body falling on it. As merely prefatory, it would be both indiscreet and unwise to enter upon the merits of this controversy; but I think it must be clear to all who have studied the question, that the entire phenomena of light are most favourable to the former view, and it would not be difficult to determine to which side the balance of probability would incline.

It appears, from the history of creation recorded in the Genesis, that light was summoned into being, in relation to our globe, at a very early period of the hexäemeron, and it may have been even localized in the orb of the sun, to attest the " evening and the morning," described by the diurnal revolution of the earth on its axis. It is true, it was not until the *fourth* day that the sun and moon were visible from the earth, and then the alternation of day and night, and vicissitude of the seasons, were established as they now stand; because, until the firmament was circumfused around the globe, and the separation of the waters took place, the earth must have been enveloped in a dense shroud of vapour, which the solar ray could not pierce. Besides all this, prior to the creation of vegetation, there could be no such thing as "*seasons*," because these periodic vicissitudes refer to vegetation, &c. and this connection alone constitutes them such. In like manner, at this early period of creation there was no animated form to which " days and years" could properly apply as a measure of age; accordingly, it was not until " the earth brought forth grass," and under the tact of the omnipotent fiat, became instinct with the glow of botanical glory, that these measures of time were introduced into the system. But light might have been otherwise localized or diffused on the first day of creation, to witness the

sublime progress of creative omnipotence. It
might have been scattered over the confused elements of chaos, to promote their organization
and separation. In a localized form, light
might have risen over the axal revolution of
the earth, and described the terminal line of
light and shade in periodic times—and it might
not have been transferred to the splendid central station it now occupies till the fourth diurnal revolution; and therefore it is evident, that
though the earth might, on the first day, have
begun to move on its axis, and the evening and
morning, in relation to this revolution, be thus
described, it might not have begun to move in
the plane of the ecliptic till the fourth day of
this sublime creative order, when the transfer-
ence of light took place; and then we are told,
by the sacred Historian, that the luminaries of
heaven, " the greater to rule by day and the
lesser to rule by night," divided the day from
the night, and were "for signs and for seasons,
and for days and for years." The firmament
had now been established, and " divided the
waters which were under the firmament from

the waters which were above the firmament."
The transfer of light to its station in the centre
of the system had taken place; and our world,
which before had only moved on its axis, began
its march in the plane of the ecliptic, obedient
to the laws discovered by Kepler, and stipu-
lated by the combined powers of the attraction
of gravitation, and a projectile force perpendicular to that of the sun. " The sun and moon
were now lights in the firmament, to give light
upon the earth,"—and the alternations of day
and night, and vicissitude of the seasons, were
firmly established. Days and years, and signs
and seasons, were now provided for. When
creation stood a finished monument of creative
power, wisdom, and goodness, and had emerged
from the fiat of omnipotence in all the " ma-
jesty of loveliness,"—we are told " the morning
stars sang together,"—perhaps this magnificent
expression is something more than orientalism.
It may be thus understood:—when the earth
took its station in the sky—no discord was
introduced into the celestial systems—the stars
still moved isochronous in their orbits, describ-

ing the lines assigned to them, and, agreeable to the laws of Kepler, preserving equal eras in equal times, which find their measure and expression in the vibration of a musical string.

Well might the " stars sing together," as they rose over the morning of a finished creation—and well might those who witnessed the glorious scene, and all the grandeur and beauty of its moving machinery, echo in loud acclaim from their celestial stations—" It is finished."

The sources of light are very various. Light is copiously dispensed by the sun and in the reflected moon-beam. Each star has its twinkling light—each planet of the system its borrowed ray. But light may be produced or elicited, not immediately or directly connected with the solar ray. The luminosity of the sea seems connected with the presence of luminous insects—as, the cancer fulgens, limulus, &c. Electric exhibitions have often the accompaniment of light, and even the *torpedo*, when touched by metallic conductors, and the circle is interrupted, will emit a spark of light. Some varieties of fluor spar emit light when heated,

as when powdered fluate of lime is thrown on
a heated plate, or projected on a surface of hot
oil of olives. The *chlorophane*, even by the heat
of the hand, yields a fine green light. Dr.
Brewster has given us an interesting paper on
the Phosphorescence of Minerals by Heat.*
Thus, the petalite is *blue* and very bright—the
green telesie is also *blue* and pretty bright—
compact fluor is a fine *green*—phosphate of lime
is *yellow*—arragonite and harmotome, *reddish
yellow*—rubelite, *scarlet. Tungstate of lime* was
brilliant like a burning coal, and *anatase* seems
peculiar, appearing suddenly like a flame, with
speedy extinction. Mr. Skrimshire has given
us a list of substances that emit light on being
brought within the circuit of the electric current
—as alum, sugar, chalk, &c. The curious
phenomena connected with phosphorescence
seems to have been first described by Benve-
nuto Cellini about the beginning of the 16th
century; and, in 1663, the Hon. Mr. Boyle

---

* Dr. B. has also favoured us with a list of several fluids
that become luminous when heated.

observed, that when the diamond was slightly heated, rubbed, or compressed, it emitted light. Some diamonds emit light in darkness—Canton's phosphorus, and the Bolognian stone, are phosphorescent when heated; and I find that a very beautiful phenomenon is presented if powdered Canton's phosphorus be strewed over a surface of mercury, and touched by the metallic poles of the voltaic circle—it becomes most beautifully luminous. Decayed wood, the potatoe, the " tremella meteorica," and some fish, as the *mackerel*, *whiting*, &c. exhibit light. Several plants emit light occasionally—a lambent light is seen sometimes to flicker on the Indian cress. Even in mines, plants, particularly those of the rhizomorpha, are found to be luminous. Mr. James Ryan* informed me, some years ago, that he had met with luminous plants in mines. The counsellor of mines, Mr. Erdmann, thus describes the luminosity of the

---

* This Gentleman received £100, and the Gold Medal from the Society of Arts, for his new method of ventilating Coal Mines, by insulating, &c. the coal field.

rhizomorpha, in one of the coal mines near
Dresden:—" I saw the luminous plants here
in wonderful beauty; the impression produced
by the spectacle, I shall never forget. It appeared, on descending into the mine, as if we
were entering an enchanted castle; the abun-
dance of these plants was so great, that the roof,
walls, and pillars, were entirely covered with
them, and the beautiful light they cast around
almost dazzled the eye. The light they give
out is like faint moonshine, so that two persons
near to each other could readily distinguish the
outlines of their bodies. The light appears to
be most considerable when the temperature of
the mines is comparatively high."

Percussion and friction are common sources
of light. When fulminating mercury is struck
by a hammer on an anvil, light is evolved.
The simple fracture of a lump of sugar in the
dark, gives out light—the collision of flint and
steel, or an alloy of iron and antimony on steel
—fragments of quartz—the rattan cane, &c.
Light was developed in the Fall of the Glacier
of Weisshorn, and during the fall of the trees in

the Slide of Alpnach, and is emitted occasionally in crystallization. Buchner, of Mayence, has shown that benzoic acid and acetate of potassa emit light during crystallization. Herman observed light to be given out in the crystallization of sulphate of cobalt, and also in that of fluate of soda. Wahler mentions some remarkable phenomena of this kind in the case of sulphate of soda. I have very frequently noticed brilliant little stars of light to pervade the glass retort in the fusion of chlorate of potassa — used for the purpose of obtaining oxygen.

Light is oftentimes the offspring of chemical changes, and is exhibited whether these mutations be the consequence of an increased density and combination, or an expansion of volume and separation of elements—so that change of volume, or combination and decomposition, are thus equally affected. When we bring equal volumes of chlorine and ammoniacal gas into contact, a light flashes in the cylinders—or even strong ammonia, when introduced into chlorine, under particular circumstances, emits a fine light and flame. Powdered antimony and arsenic are thus ignited—copper burns with flame, and also mercury when on the point of ebullition. The ignition of platinum wire in the vapour of ether, the aphlogistic lamp, and modification of this phenomenon discovered by Mr. Edmond Davy—and still more curious and extraordinary ignition of platinum, as shown by Dobreiner of Jena, are all singular exhibitions of light connected with chemical phenomena and the process of combustion. Intense light is given out in the burning of olifiant gas—in the transit of bubbles of per-phoretted hydrogen into oxygen—in the combustion of phosphorus in oxygen, nitrous gas and nitrous oxyde—and intense and beautiful is the light evolved between the accuminated points of well-burnt boxwood charcoal, when forming the terminations of the conducting rods in the voltaic circle, and brought into contact. A very brilliant light is also exhibited when deutoxyde of barium is heated by the spirit lamp in pure hydrogen, dissicated by transit through hydrochlorate of lime, and resting over mercury. The ignited gases from the compound gas blow-pipe, when directed on magnesia, &c. exhibits an intense and very beautiful light.

When air is compressed in the condensing syringe, if the cylinder be of glass, as Mon. Biot has shown, light pervades it; and even, according to Dessaignes, water, when compressed, is luminous. These phenomena are connected with increase of density. On the other hand, if an air-gun be discharged in the dark, a lambent light may be seen at its orifice, though this, by some more recent experiments, would seem to be the consequence of silicious matter being present. When balls of oxygen are broken in vacuo, and the separated elements expand in the explosion of chloride and iodide of azote; and also when we expose nitrate of ammonia to a specific temperature, it is decomposed with the evolution of a beautifully coloured flame—at the close, too, of the obtainment of nitrous oxyde from the nitrate of ammonia, when the temperature is raised above 500° Fahrenheit, the retort is instantly filled

21

with a yellow light, accompanied by slight explosion. It is obvious, that all these exhibitions may be resolved into chemical changes, and that the change of volume either way does not influence the evolution of light. Generally, the accompaniment of light may be considered dependent on that chemical process we call combustion, as in the light from potassium heated in a medium of the vapour of iodine. In the combination of sulphur with copper, lead, potassium, sodium, &c. or the alloys from antimony, zinc, bismuth, tin, &c. with platinum—and the more fierce and energetic the chemical action and combination, the more intense and brilliant will be the light emitted.

It is probable that a certain degree of temperature, in every case, will be found necessary for the evolution of light. This is the case with the *chlorophane* and others, and even with the rhizomorpha in mines, and, as I have found, it is the case also with the luminous matter of the glow-worm—while Dr. Ure shows that a low temperature suspends the light emitted in the case of the whiting, &c.—a higher tempera-

c

ture decomposes the luminous matter, and of course extinguishes that light which is the necessary attendant and accompaniment of a peculiar organized structure. Even a very considerable degree of cold does not unhinge the organization of many insects, &c. which may, by a carefully applied warmth, revive and become animated, though they may have been for a considerable time in a frozen state. These facts show clearly that the light in luminous animals is a consequence of, and has its being in, a peculiar organization. This conclusion is further corroborated from the experiments I have made with the luminous matter of the glow-worm, for when I ruptured the capsule which contained the light, and thus destroyed its structure, the light became evanescent.

The light of the sun, and electric light, must be regarded quite distinct from that connected with chemical combustion;—and that of the animal system, in the case of luminous animals, must also be considered apart from any thing like combustion. In the case of the luminous spherulæ of the glow-worm, it is imprisoned in a sac or transparent capsule, through which, as through a window, the light shines.

Solar light has chemical powers unpossessed by any terrestrial light. Equal parts of carbonic oxyde and of chlorine, when posited in the solar ray, Dr. Davy found, were condensed into half their volume, and gave rise to a new and peculiar gas, called by its discoverer *phosgene gas*, being formed by the exclusive agency of light. In like manner, equal volumes of hydrogen and chlorine, when thrown up in the sunbeam, combine with light and explosion, and form muriatic acid gas. Now, Professor Brande has made the capital and important discovery, that the light evolved between charcoal points in the galvanic battery, promotes the combination of chlorine and hydrogen, or formation of muriatic acid gas from its constituents—and there is thus drawn between solar and electric light the lines of a singular and close analogy —a conclusion, indeed, to which the genius of Sir Humphry Davy had previously conducted him.

When a beam of solar light is intercepted by

a prism, it unfolds a very beautiful and interesting spectacle—seven colours are presented,
and the " bow of promise in the storm" affords
a magnificent illustration of its phenomena.
The colours thus unveiled are violet, indigo,
blue, green, yellow, orange, and red, and they
are refrangible in this order, the violet being
the most easily refracted or bent, and the red
with difficulty refracted. Agreeable to the doctrine of chromatics, taught us by Sir Isaac
Newton, when all these colours are absorbed,
the object is black. Black, therefore, is the
absence of all colour. On the other hand, when
all the colours are returned or reflected, the
object is white—when a particular colour is
reflected while the others are absorbed, the
body appears of that particular tint. Thus the
chromate and hydriodate of lead reflect the
yellow ray, and therefore appear to be yellow;
and the hydriodate and persulphuret of mercury reflect the red ray, and thus appear in a
vesture of scarlet.

That colour is the gift of light, numerous
phenomena concur to assure us. The play of

colours which supervenes on changing the angle
of vision is of this description, as the dichroite,
the fire, or precious opal, the Lumachella marble, the feldspar adularia, and that from the
coast of Labrador. Entomology furnishes numerous proofs, as in the dermestes violaceus,
chrysis fulgida, libellula varia, melolontha farin-
osa, curculio imperialis; and in other insects,
as in particular species of carabus, buprestis,
and chrysomela. In the ornithological department of zoology, we have additional evidence
of this beautiful mockery of vision, as in the
" alcedo cristata," several of the species of trochilus, or humming bird, &c.; add to these the
change of tint in the " indigo bird," described
in Mr. Wilson's American Ornithology.* It
appears at one time of a rich sky-blue—at
another of a vivid verdigris green—so that the
same bird, in passing from one place to another
before your eyes, seems to undergo a total
change of colour. When the angle of incidence,
in the rays of light reflected from the plumage

* Vol. I. p. 101.

c 2

of the bird, is acute, the colour is green—when obtuse, blue. The colour of the head being a very deep blue, is not affected by any change of position. Prince Maximillian, in his Travels in the Brazils, says, that in the procnia cyantropus held against the light, the body seems wholly a splendid sky-blue, if turned from it, then it appears a shining light-green.

Notwithstanding that the doctrine of colours appears thus easily deducible, there remain many circumstances extremely difficult to be accounted for. Colours are thus to be considered accidental rather than essential properties of bodies. Sir Isaac Newton fills up the solar-spectrum with *seven* colours, and he found that, by collecting them together by means of a convex lens, they formed a spot of bright white light. This eminent individual found no decomposition of either the orange or green colour of the spectrum, while the prism soon resolved orange or green, formed by mixing the yellow and red, or yellow and blue, of the solar beam thus decomposed, into their respective elements. To this apparently *experimentum crucis*, it might be sufficient to reply, that it could scarcely be expected that a prism could undo that which itself had formed. Dr. Wollaston reduces the seven colours into *four*. This philosopher, by receiving a narrow line of light on a prism, had only presented to him four colours; namely, red, green, blue, and violet;—a very narrow line of yellow was visible at the termination of the red and green, which Dr. Wollaston attributed to the overlaying of the edges of the red and green rays. But surely, *prima facie*, an admixture of red and green could never produce yellow, nor will the one, superimposed on the other, exhibit this tint.

Dr. Wollaston observed, that by increasing the breadth of the opening which admitted the line of light, the space occupied by each coloured ray in the spectrum was augmented in proportion as each encroached on the neighbouring one and mixed with it, so that the intermixture of orange and yellow between the red and green, he presumes to be owing to the mingling together of these two colours, and that the blue is blended on one side with the green, and on the other with the violet, forming the

spectrum as originally observed by Sir Isaac
Newton. This, it must be confessed, is not
only ingenious, but original and profound, like
every thing that comes from this distinguished
philosopher—while, on the other hand, Dr.
Brewster contends strenuously for the yellow
ray having a distinct and independent place in
the spectrum.

The existence of dark and bright lines traversing the spectrum, was determined by Dr.
Wollaston.* Dr. Young theorized on the phenomena.† M. Frauenhofer has described their
position minutely, and has also used them as
definite points for the measurement of the dis-
persion. The prismatic spectrum also possesses chemical agencies or powers—the violet
ray terminating the spectrum on one hand,
separating oxygen and combining hydrogen,
and the red ray, on the other, separating hydrogen and combining oxygen. These curious
and interesting facts, pointed out by Wollaston,

* Phil. Trans. 1802, pp. 378 and 380.
† Ibid. p. 395.

Davy, Ritter, and Boëkman, have been confirmed by Berard.*

Sir Wm. Herschell found that the coloured rays of the spectrum possessed different heating powers; a delicate thermometer, placed in the violet ray, rose 2^0 Fahrenheit. The average of the green amounted to $2^0.25$, and of the red $4^0.5833$. Sir Henry Englefield and M. Berard have confirmed these results.

Mr. Baden Powell, of Oriel College, Oxford, in a recent communication to the Royal Society of London, has described the advance of a differential thermometer, with a blackened ball, to be from 16^0 to 23^0 in the extreme, visible red; while the effect exterior to the red rays was 3^0; when merely blackened, it was 12^0

---

* The amount of analogy between solar and electric light is thus exalted to a maximum. The new phenomena of electro-magnetism, when viewed in connection with the magnetism imparted to delicate steel bars by violet solar light, by Professor Morrichini of Rome, closes the probabilities, or swells them almost to positive conclusion. In 1818, Signor Morrichini, while he presented me with one of the needles so magnetized, informed me that from some new and interesting experiments, he had found the solar spectrum a very delicate galvanic battery.

in the red ray, and 2⁰ beyond its boundary—
while covered with thin yellowish brown silk,
it stood in the red at 7⁰, and half-an-inch
beyond at 7⁰. Beyond the red verge the inter-
position of glass diminished the effect, and
hence Mr. Powell has concluded, that " the red
rays radiate outward, and from their own particles." These effects, he has added, are more
developed by a thicker and rougher coating
than by being merely blackened, or from darkness of colour.

In experiments such as these, it seems to me
most clear that sufficient attention has not been
paid to the colour or texture, and other properties of the material which enveloped the ball
of the thermometer. In order to exemplify my
meaning, I would just state, that if I grasp in
the palm of my hand a delicate thermometer,
the instrument has indicated 98⁰.5 Fahrenheit;
whereas, when I have loosely or in a spongy
form enveloped the ball of the thermometer
with copper-foil or silver-leaf, the instrument
has mounted up to 101⁰ Fahrenheit. If we
grasp either copper or silver-foil in the hand,

we feel a sensible glow of warmth. It is perhaps difficult, from our present knowledge, to
explain this phenomenon, and I therefore
merely register it as a fact.

The temperature of colours has engaged
much of my attention, and in a Pamphlet on
the Physiology of the Corolla of the Flower,
and in a communication made through Dr.
Brewster to the Royal Society of Edinburgh,
and now deposited in their *archives*, I have
detailed my experimental researches connected
with a question I believe entirely new; yet, if I
mistake not, promising to throw considerable
light on a curious subject, and susceptible of
much practical utility. I find a singular uniformity whether I place the ball of a delicate
thermometer on the coloured corolla of the
flower, or plunge it into the new-formed colour, the result of chemical change—and I have
concluded from hence, that temperature be-
longs, as an essential attribute, to colour, and
is inherent in it as a characteristic feature, the
increments of temperature being in some definite ratio of the brilliancy of the light which

beams upon the colour, and from which it receives its grade of intensity. I have, therefore, concluded the production of colour in chemical changes to be accompanied on the instant by a temperature indigenous, or peculiar to that colour, and corresponding to the same tint as exhibited in the prismatic series.

In the case of white formed from subnitrate of bismuth, when the temperature of the air was 47° Fahrenheit, I obtained only the fraction of a degree more; while in *black*, resulting from equal portions of nitrate of iron in solution and tincture of galls, I had 9° Fahrenheit above the ambient air, and 8° above the average temperature. In *blue*, from nitrate of iron and ferrocyanate of ammonia, the increment was 1°.5 above the air; in the chromate of lead, a brilliant yellow, produced from acetate of lead and chromate of potassa in solution, the acquired temperature was 2°.5; while that produced in the dazzling red of the hydriodate of mercury was 7°.5—the change of density, it was proved, had no share in the phenomenon of temperature, and that which might be sup-

posed connected with the chemical action superinduced, was always suffered to subside—while
the ambient air was 79⁰, the petals of the
white lily exhibited the same temperature—the
air being 77⁰, the blue tradescantia was 79⁰;
while the air was 76⁰, the central yellow base,
or tongue of the gum cistus, was 79⁰; and the
air being 81⁰, the scarlet geranium was 87⁰.
It was interesting to note the various changes
which supervened, while I on one occasion
made my experiments on the red flower of the
flos adonis—the thermometer fell several degrees on the transit of a cloud over the solar
disc.

On the 31st December last, at 45 minutes
past 11 o'clock, A. M., I made some experiments
on the *mineral chameleon*. Air was 45⁰.5; water
40⁰.75; a little of the mineral chameleon, diffused through the water, communicated an
emerald green tint, which became speedily blue,
and the thermometer, at first 41⁰, sunk to
40⁰.25; with a few drops of caustic potassa, a
dark green was formed, and the thermometer
stood at 41⁰.25; a little nitric acid added, pro-

D

duced a port wine or beet red colour, and the thermometer rose to $43^0.75$. These experiments were repeated three times. In the second series, the thermometer from emerald green to blue fell from 41^0 to 40^0; in the dark green, from alkali, the thermometer stood permanent at $42^0.5$. In the third series, water was 41^0; and the instrument, in the transition from green to blue, fell from $41^0.75$ to $41^0.5$; with alkali, it rose to $42^0.25$ (dark green), and with a drop of acid, producing a dark purplish red, a temperature of $43^0.25$ was formed.

My experiments on the temperature of colours go to prove that there are only *three* colours in the spectrum; namely, blue, yellow, and red. *Carmine*, and not vermilion, is the best representation of this last in the series of primitive colours. Black seems to be a *compound* colour, composed of all the three colours, while white is the result of their entire absence. This view perfectly harmonizes with the opinion of Mr. Hargreaves and others. I find, *experimentally*, green and orange to be the added temperatures of blue and yellow in the former;

and yellow and red in the latter: black, the added temperatures of blue, yellow, and red. The artist can make black by mixing together Antwerp blue, gamboge, and carmine. We know that the finest black cloth is first dyed blue: and Dr. Bancroft informs us, that the superior blacks of the once famous Gobelins of France, were produced by passing the cloth through woad, weld, and madder, or blue, yellow, and red.

By a delicate thermometer, the artist and manufacturer may hereafter attain a superior accuracy in the composition of their colours and dyes; nor is it even too much to hope yet to see the painter avail himself of this important instrument in the comparison of the colours of his canvas with the rich and beauteous tints of creation.

The singular disposition of the colours in the tabernacle in the wilderness, as well as the temple at Jerusalem, and the unvarying order in which they are ever mentioned, in this relation, in the sacred volume, is something more, as it occurs to me, than a happy incident or

remarkable coincidence—the " blue, purple,
and scarlet," the two limiting powers of the
spectrum, mingling centrally into one. This
seems unquestionably to symbolize the " bow
in the cloud," which appeared on the recession
of the diluvial waters;—emblem of the divine
beneficence to man—

> ——————" Like the bow,
> Called out of rain-clouds, hue by hue—
> Saw the grand gradual picture grow,—
> The covenant with humankind
> Which God hath made." *

The very covering of the sacred ornaments
bears the same relation:—" Goats' skins dyed
red, and *hyacinthine*, or violet skins," as it might
more properly be translated from the Hebrew.
The badger was unknown in Palestine, and
even the Greeks and Romans had not a name
by which to express it. Besides, it was an
unclean animal, and consequently interdicted
by the Mosaic law; and, therefore, could not

---

* The Loves of the Angels. T. Moore. London. Third
Edition. 1823. P. 69.

be a proper covering for the sacred and costly ornaments of the tabernacle. In this view, there is something extremely interesting and apposite in the disposition and arrangement of the colours—nor is it too much to suppose that these might have both an *isoteric* and *exoteric* meaning; and, though the Jewish lawgiver might be even himself ignorant of the full amount of the philosophy included in it, it must be ever remembered, that the whole was the order assigned by Jehovah, and according to " the pattern showed in the mount." It was the interesting pledge, that " cold and heat," " summer and winter," were not henceforth to cease; and are not the temperatures of the blue and red admirable symbols of the difference of temperature in summer and in winter ?

These preliminaries, on the question of light generally, cannot be inapposite to the consideration of it as connected with the animal system, in which it seems so strange and peculiar a feature—and it is perhaps partly from considering light as thus connected, that we shall be able to collect new facts toward its elucidation.

Of luminous insects in this country, the lampyris noctiluca, or glow-worm, and the scolopendra electrica, are the most conspicuous and common. The scolopendra electrica I first found in Huntingdonshire, and the lampyris splendidula most abundantly in the vicinity of Oswestry, which indeed seems a prolific region for them. It had been supposed, that the property of emitting light was confined to the female in the lampyris noctiluca. Ray first pointed out that the male insect too yielded light, and the circumstance was confirmed by Geoffroy and Müller. In the males of lampyris splendidula, and lampyris hemiptera, the light is distinct, and may be recognized when the insect is on the wing. Still, the light is of inferior brilliancy, and confined to minute points. But the glow-worm, in the case of the larva and perfect insect, have different degrees of luminosity, and sometimes may have been confounded together. It has been observed, that the females of the glow-worm can occa-

sionally conceal or eclipse their light—and it
may be to secure themselves from becoming
the prey of the nightingale or other nocturnal
birds. Mr. White, indeed, thought that they
regularly extinguished their torch between the
hours of eleven and twelve. The light with
which it is invested, may perhaps occasionally
deter its insect enemies from making an attack
on it—as the wolf, and other ravenous beasts
of prey, are deterred from making an approach
on travellers by night when encircled by fire.
Though, it must be admitted, that many of the
species of the lampyris are without wings, and
want even elytra—the female of the *lampyris
italica* is a winged insect. Mr. Waller, in the
Phil. Trans. for 1684, informs us he caught
one in Hertfordshire. I can remember to have
seen the Spanish fly (*cantharis*) in a garden in
the same county, namely, at Broxbourn. The
lampyris italica I used to collect in Tuscany,
near to Leghorn, along with the late Mr.
Brown, the British Vice-Consul at that port.
Sir James E. Smith, in his Continental Tour,
informs us that the beaus of Italy oftentimes

adorn the head-dresses of the belles with these
" stars of the earth and diamonds of the night."
A similar practice prevails in India. Dr. Clau-
dius Buchanan, in his " Christian Researches,"
mentions having written one of his letters under
the pendant nest of an oriole bird, which had
lighted up its abode with a luminous insect—
" as if," the Doctor facetiously remarks, " to
see company." The Italians have a super-
stitious dread of these beautifully adorned
insects ; believing them to be the spirits of their
departed ancestors. Sir J. E. Smith mentions
that some Moorish ladies were taken prisoners
at sea, and until they could be ransomed, lived
in a house beyond the city walls of Genoa :
during the period of their stay, they were visited
by respectable inhabitants of the city :—on one
occasion of their visit, they found the house
shut up, and the Moorish females in great grief
and consternation. Some of the beautiful lampyris italica had found ingress into the house,
and these brilliant guests were supposed to be
no other than the troubled spirits of their rela-
tions. I remember, one fine night, on coming

from Arcqua, (where I had been to see the tomb and former residence of Petrarch,) to Padua, that the whole trees and hedges, to the very summit, were illuminated with myriads of these living diamonds—the effect was magically magnificent.

The genus lampyris is not exclusively luminous—the elater noctilucus is invested with light, even in a superior form. The light in this curious insect is emitted from two protuberant, transparent, or windowed tubercles attached to its thorax; besides these, however, there are two luminous spots beneath the elytra, only visible, of course, when it is on wing, and they are elevated; it then appears studded with four rich and vivid gems of a golden, blue lustre. In fact, the whole body seems a flood of pure light. In the West Indies, particularly St. Domingo, the natives employ these living fires to give light in managing their household concerns. In travelling, they are wont to attach one to each toe; and it is stated that in fishing and hunting they require no other illumination.

Pietro Martire informs us, that the elater noctilucus serve the natives of the Spanish West India Islands both for lights to lighten their dwelling, and to extirpate the gnats. Introducing the fire-flies into their houses, the gnats become their prey; and, if they could have a like power over that pest, the *mosquito*, it would be worth while to introduce them into some parts of Italy, and to cultivate and naturalize them. The pain and irritation I have experienced from mosquitoes at Venice, would make me esteem no sacrifice too great to get rid of them. The lampyris italica which I perceived on the wing on the road to Padua, might have diminished that plague, which could find no equivalent for me in Italian scene, clime, and sky, enthusiastically though I be attached to the land of song and classic beauty. On festival days, these fire-flies are collected and attached to the clothes, and the horses: and, according to the same author, the luminous matter is rubbed over the face occasionally, as phosphorized oil is used. Elater ignitus is another luminous species of the same genus.

We are told by Mouffet, that the appearance of the tropical fire-flies, on one occasion, led to a singular result. When Sir Thomas Cavendish and Sir Robert Dudley first landed in the West Indies, the flitting and moving lights of these insects in the woods gave them an impression of the advance of the Spaniards, and they returned in consequence to their ships. The genus *fulgora*, (order hemiptera) or lantern-fly, emits a powerful light. The fulgora lanternaria is a native of South America, and the fulgora candelaria a native of China. The light in these species is included in a hollow subtransparent projection of the head, a kind of proboscis. Splendid, indeed, must be the appearance of a tree thus singularly adorned and studded by so many sparkling stars. In the fulgora lanternaria, the light perhaps transcends that of any other luminous insect. The light, according to Madame Marian, is sufficient to read a newspaper by. I have myself read a letter by the exclusive light of the lampyris noctiluca, and last summer I alighted at night on the Italian side of the Simplon, near to the white marble quarry, and picked up a lampyris splendidula. It shone with intense light; the night was dark, but I preserved the insect, and carrying it into the carriage with me, turned it to good account, for it showed me distinctly the hour by a watch. The scolopendra electrica and phosphorea are two species belonging also to hemipterous insects, and yielding light. It has been inferred, by Messrs. Kirby and Spence,* that some insects may occasionally exhibit light, not commonly suspected—and the mole cricket, *gryllotalpa vulg.*, is cited, on the authority of the Rev. Dr. Sutton; and it is even added, that the *ignis fatuus*, or Jack o' Lantern, may be of this description. Dr. Darwin, I am informed by one of his friends and acquaintances, never would believe in the existence of the ignis fatuus. It was, therefore, expunged from his creed. He had never seen it—this was enough for his scepticism—which included, on similar principles, revealed reli-

---

* Introduction to Entomology. London. 8vo. 1817. Vol. II. p. 421.

gion; yet, strange inconsistency, Dr. Darwin was marvellously credible in many other matters. I never myself witnessed this curious phenomenon but once—when riding at night between Glenluce and Newton-Stewart. It appeared as a glow of lambent flame emerging from the surface of the lake. It had no semblance to the flitting motion of an insect, and it as little resembled perphosphoretted hydrogen.

The following observations on the glow-worm are by " W. Rogerson," and are addressed to the Editor of the Philosophical Magazine.*
" The female deposits her eggs in the month of June or July, among moss, grass, &c. These eggs are of a yellow colour, and emit light. After remaining about five or six weeks, the larvæ break their shells and make their appearance; at first they appear white, and are very small, but they soon increase in size, and their colour changes to a dark brown, or nearly black colour. The body of the larva is formed of eleven rings. It has six feet, and two rows

* Phil. Mag. Vol. LVIII. p. 53.

E

of reddish spots down the back. It emits light
in the dark; this light arises from the last ring
of its body under the tail, and appears like two
brilliant spots when attentively examined. The
larvæ are seen creeping about and shining
during the fine nights in autumn, and the light
they emit is to direct them to their food. They
feed on small snails, the carcases of insects, &c.
They frequently cast off their skins: after the
expiration of about one year and nine months
from their birth, they arrive at their perfect
size. They then cease to eat, cast off their
skin, and assume another appearance. The
form of the perfect insect may be discovered
through a thin skin that covers them. After
remaining in this state two or three weeks,
(scarcely ever moving,) they throw off their last
skin, and arrive at perfection. The male then
appears a perfect beetle, having wings and
covers to the same. The female, on the contrary, has neither wings nor wing-cases; she is
larger than the male, and of a lighter colour.
It is the female that principally shines in a
perfect state. Her light is far superior to that

emitted by the larvæ, and arises from the three last rings of the body on the lower side."

Numerous opinions have obtained on the proximate cause or source of this curious illumination. Mr. Macartney, in the Transactions of the Royal Society, has given an interesting detail of experiments and observations on several luminous insects. In the lampyris noctiluca, and in the elater noctilucus and ignitus, the light, he observes, proceeds from a substance not distinguishable from the interstitial substance except in its colour, which is yellow. The light-yielding matter reposes underneath those transparent parts of the skin, through which casement it is seen. He infers that the luminous matter in the glow-worm is absorbed, being replaced by the interstitial matter when the season for emitting light is gone by. He observed two minute elliptical sacs formed of an elastic fibre, wound spirally and similar to that of the tracheœ, which contained a yellow substance, soft in consistency, and closer in texture than that lining the adjoining region, and affording a more brilliant and permanent light. This

light he concluded to be less under the control
of the insect than the luminous substance in its
vicinity, which he inferred it had the property voluntarily to extinguish, referable to
some inscrutable power dependent on volition,
and not, as was advocated by Signor Carradori, by retracting it under a membrane—when
he extracted the latter from living glow-worms,
it afforded no light, while the two sacs in like
circumstances shone uninterruptedly for several
hours. The reason, however, of the apparent
extinction of the luminous matter in the penultimate and anti-penultimate annulæ, I have
found, is the envelopement of the luminous
matter by the surrounding interstitial mass,
and if it be carefully sought for, it will be found
deeply imbedded in it: this is the reason of the
eclipse. Mr. Rogerson says the eggs of the
insect are luminous. I have certainly found
luminous matter excreted from the glow-worm,
and this luminous matter thus occasionally
displayed and oftentimes submerged and lost
in the common interstitial mass, may be the
ovarium ; at any rate, when the luminous mat-

ter is found, it appears under the form of
minute brilliant points. Mr. Macartney is of
opinion, that the interstitial substance which
surrounds the oval yellow masses under the
transparent spots in the thorax of the *elater
noctilucus,* has the power of exhibiting light; in
which inference he considers himself warranted
from the radiated structure of this substance
conjoined with the translucency of the adjoining
crust. Mr. Macartney was unable to conclude
upon the peculiar organization that contributed
to the efflux of light in the hollow projections of
fulgora lanternaria and candelaria—the hollow
antennæ of pausus sphœrocerus, and that under
the entire integument of the scolopendra electrica. Respecting the scolopendra electrica, he
conceives that it will not shine unless it be previously exposed to solar light—this conclusion
is, however, not warranted by the premises.
Mr. Macartney has even in other cases inferred
that solar light is injurious to luminous animals,
and that in the case of one at least of those in
the sea;—it retreats from the surface. He
tried electrical stimuli, but supposes it acted

merely as a mechanical stimulus in educing light.

Dr. Darwin has presumed that the luminous exhibition was owing to a secretion of some phosphoric matter, and was a slow combustion arising from this matter of phosphorus entering into combination with the oxygen inspired— the large spiracula in glow-worms seemed to give a plausible colouring to the idea. It was also stated that the light was increased by heat and oxygen—and extinguished by cold and hydrogen, and carbonic acid gas. I have clearly ascertained, however, experimentally, that the luminous matter does *not* contain phosphorus —and that the light is not sensibly increased by the purest oxygen—and is *not* extinguished in hydrogen and carbonic acid gas. Spallanzani regarded the luminous matter as a compound of hydrogen and phosphoretted hydrogen. Carradori found that the luminous matter of the lampyris italica shone in vacuo, oil, water, and under other circumstances, to the exclusion of oxygen; and he and Brugnatelli concluded, that the emission of light was owing to its absorp-

tion in food or air, and subsequently secreted in a
sensible form. Mr. Macartney ascertained that
the light of the glow-worm is not diminished
by immersion in water, or increased by the
application of heat. I find, experimentally, that
the maximum of brilliancy is exhibited at 98° or
99° Fahrenheit, and that it declines as the heat
advances. Foster and Spallanzani assert that
glow-worms shine more brilliantly in oxygen;
while Beckerheim, Dr. Hulme, and Sir H.
Davy, discovered no such effect—and to these
last I adjoin my testimony. Spallanzani and
Dr. Hulme state that the light of the glow-
worm is extinguished in hydrogen and carbonic
acid gas. Sir H. Davy, on the other hand,
found that this light was not sensibly diminished in hydrogen, and I have not only found
the same thing, but that the animal itself does
not seem to suffer materially in this gas; and
that though it expires in carbonic acid gas, the
light suffers no eclipse by the fate of the insect.
Dr. John Davy found, at Kandy, on 30th June,
while the air was 73°, that the temperature of
a large species of glow-worm was 74° Fahren-

heit; and I have found my thermometer affected by nearly a degree from suffering the glow-worm to crawl over its bulb. I purposely omit any mention of Dr. Todd's paper on the glow-worm, &c. transmitted to the Royal Society of London, because my communication, on the same subject, had been, a twelvemonth before, in possession of the Linnean Society: and an abridged account of it had appeared some months before in the Philosophical Magazine, and in other journals, nor do I find he has added one fact to the subject. My *experiments* are more extensive at least; though I have, for obvious reasons, been more cautious in deductions, and sporting with speculations, which have generally only a playful fancy to recommend them to notice or attention.

On the night of the 10th of June, 1822, on returning from Llanymennech to Oswestry, in Shropshire, I picked up two glow-worms from the grass on the roadside. The brilliancy was intensely beautiful—the penultimate and anti-penultimate rings of the under side of the abdomen were the portions illuminated, together with two luminous points on the anal segment. The night was dry, and much lightning obtained: I repeatedly held the insects exposed and attached to the spikes of grass—to the repeated flashes of lightning which played on them, but this seemed by no means to disconcert them; their luminous paraphernalia were still unveiled. One placed on the watch-glass gave sufficient light by which to tell the hour. I arrived at half-past 10, P.M., and introducing them into a room with a candle, the light gradually diminished in intensity, and became ultimately extinct, attenuating from the edges of the bands or discs, and disappearing in patches toward the centre. The insects, when undisturbed, seemed ever anxious to gain the summits of the blades of grass toward evening and at night, and from thence, as from a Pharos, displayed their beauteous insignia by holding up the lower rings of the abdomen. When they had ceased to shine, and during the day, they sought the lower parts of the box or grass. Thermometer stood at 70° Fahr. When the luminous matter was carefully in-

spected by a lens, not the slightest undulatory
motion could be perceived, and the luminous
matter appeared incased in a transparent sac.
On the 12th June, at 20 minutes past 7, P. M.,
excluded from light, it had already kindled
up a faint beam. When I exposed the insect
to light, it ceased to shine. It appeared annu-
larly translucent in the membrane which
covered the luminous belts. At 50 minutes
past 9, P. M., it shone exceedingly bright, still,
by the most careful inspection, no undulatory
movement could be seen, which proves it to be
entirely unconnected with any phenomenon in
the shape of a languid combustion, and though
the light evidently intermits, it does not glow as
in aphlogistic exhibition. While it remained
attached to a spike of grass, I could easily read
a letter by the light it shed, moving the insect
thus over the lines. About this period, thermometer constantly ranging from 70° to 80°
Fahr., the glow-worm shone most vividly, and
one was so intense on the road from Oswestry
to Ruyton, that a horse started at the gleam,
and nearly overthrew its rider. My observa-

tions would certainly prove that the luminous matter is under the control or management of the insect, and I have noticed that the light is kindled up when it moves on the leaf, and that its palpi are then employed. I have found five luminous specks, of a minute oval form, in the box which contained my glow-worms; they were evidently secreted by the insect, and might be the ova. They continued luminous for some time, the light was of a different tint of colour, which I account for on the simple intervention of atmospheric air in one case, and the transparent film of the membrane which enveloped it in the other. I always found that my glow-worms concealed themselves during the day among the roots of the grass. On holding the candle near, they do not withdraw their shining immediately, but it gradually attenuates. I let one fall on a cold slab of marble, but it still shone brilliantly—the insect doubtless having the power of sustaining a superior temperature.

The physical strength of the glow-worm seems considerable. I had about two dozen in a box, supplied with grass and moss : the lid was loosely tacked down, and, as I had persuaded myself, secure enough for them. In this, however, it appears I was mistaken; for, on returning home one night, I found the lid of the box detached, and my glow-worms scattered on the carpet of the room—the floor appeared thus powdered with diamonds.

I must acknowledge here my obligations to T. N. Parker, Esq. of Sweeney Hall, through whose kindness I was abundantly supplied with these insects, for the purpose of experiment— the hedges and grass around his seat being sprinkled with a profusion of them.

After the luminous sacs were detached from the glow-worm, the light in the penultimate and anti-penultimate rings was absorbed, the two luminous points on the anal sac alone remained. But by proper pressure, &c. the luminous spherulæ might be detached entire, and if not crushed or lacerated, would continue to evolve light after their detachment from the nidus in which they were imbedded — the luminous globules are completely independent altogether of the interstitial substance through which they are diffused. The fact of the gradual and not instantaneous evolution of the light, and the gradual eclipse it suffers, seem to refer the phenomenon to the volition of the insect—the intervention of a shade or contraction of the fibres on which the luminous points are suspended, or to which they are attached, would account for the occasional occultation of the light. The luminous matter heated in a platinum spoon, soon ceased to evolve light. The surface of the platinum exhibited no change, which it would have suffered had the luminous matter contained phosphorus as one of its chemical constituents.

The light of the glow-worm seems monochromatic, and appears incapable of further decomposition by the prism. Viewed through a prism of rock crystal turned on its axis, it presented a confused nebulous image, unaccompanied with the chromatic tint. The photometer was affected by 2^0 or 3^0 on its near approach : this could only be inferred from the apparent sinking of the liquid, calculating from

F

the moment of observation to that when it
became stationary.

The luminous spherulæ secreted from the
insect, when dry, were heated over a spirit lamp
in a platinum spoon: they became dark brown,
and exhibited a momentary flame, accompanied
by slight explosion—the flame, in colour, &c.
had all the characters of hydrocarbonate. By
a delicate chemical analysis, I have inferred it
to be a gummo-albuminous substance.

Luminous spherulæ detached from the glow-
worm, were immersed in various media to
ascertain the period of continued luminosity.
In a concentrated solution of pure caustic potassa, the light was of a bluish tint, and seemed
to undulate; the period of duration was 60
seconds, and it continued luminous for several
minutes in tincture of iodine. In sulphuric acid,
of specific gravity 1.85, it was luminous for 30
seconds, and was finally blackened. In muriatic acid, specific gravity 1.12, the light
remained for 24 seconds; and in nitric acid,
specific gravity 1.45, it was only luminous 10
seconds. In a subsequent experiment with

sulphuric acid, specific gravity 1.85, it remained luminous only 20 seconds.

In alcohol, specific gravity .812, the luminosity lasted nearly two minutes; and in camphorated spirit of wine, nearly the same period. In caustic ammonia in solution, it was luminous for 60 seconds. In almond oil, and in water at 68° Fahrenheit, it continued luminous for some hours, but was much less brilliant and beautiful than when immersed in oil of olives. When put into olive oil, at 11 o'clock, P. M., it continued to yield light through the entire night, and several successive ones: when reposing in this tranquil medium, and viewed at a distance of about 10 feet, it twinkled like a fixed star, and although the eye steadily and tranquilly observed the beautiful phenomenon, the light seemed subject to occasional occultation, as if some axal revolution had been superinduced; and this curious eclipse took place in periodic times—but I could not determine whether this phenomenon was dependent on some property inherent in the light itself, or the result of circumstances connected with the

organ of vision, or the air itself. Two glow-
worms were shut up in a dark box from Tues-
day morning to Saturday night; yet, when then
examined, were shining with great brilliancy—
a fact which tends to prove that previous exposure to light is not necessary to their luminosity, and, conjoined with the circumstance that
during the day they penetrate to the roots of
the grass, or base of their stems, would lead us
to conclude, that solar light, on the contrary, is
injurious in this relation. When I breathed
on a glow-worm not previously shining, it
became luminous—probably from the accom-
panying increment of temperature.

The luminous matter continued to shine, without alteration, in oxygen, hydrogen, carbonic
acid gas, cyanogen, olifiant gas, nitrous gas, &c.

When I introduced the insect into pure
oxygen, obtained from chlorate of potassa, it
did not seem more alert than before; neither, in
repeated trials, was the luminosity perceptibly
increased. The insect was not injured by being
introduced into a medium of hydrogen, nor did
the light undergo any change. In carbonic

acid the glow-worm soon exhibited signs of suffering, and expired in a bright shroud of light it had no power to quench—the light continued a considerable time after the death of the insect. As in the Persian poem of Sadi, so the luminous matter continued to reflect its gleam on the parent of its existence; or as in the fabled song of the dying swan, reserved its brightest glow for the last throb of life.

Light does not seem to exercise any control over the luminous matter when reposing in oil of olives; indeed, a luminous spherule exposed for an entire day to the sun-beam, shone at night as usual, or with slight decrease of intensity.

A luminous spherule held in my warm hand, say 98° Fahrenheit, for two minutes, continued luminous as in ordinary circumstances. It was then introduced into oil of olives, where it shone uninterruptedly. In water heated to 108° Fahrenheit, the luminous matter continued to shine for more than a minute; in a temperature of 114° Fahrenheit, it became extinct in about 40 seconds; in 126°, it was lost in 30 seconds; in a temperature of 94°, it became *faint* in 60 seconds, though it continued to emit light much longer; in a temperature of 99° Fahrenheit, the light was more *brilliant* and *beautiful* than in any other temperature to which the luminous matter was submitted, and it continued more intense for about 60 seconds than in any other case.

A glow-worm not luminous, was introduced into water at 64°.5 Fahrenheit, but no light was emitted—at a temperature of 88°, light was faintly emitted at the last, or anal segment. In water of the temperature of 110° Fahrenheit, it crawled about apparently disconcerted, emitting two gleams, which were as suddenly quenched : but on being withdrawn, it did not seem inconvenienced. The insect died at a temperature of 125° Fahrenheit; but being suffered to remain in water at 116° Fahrenheit, light was educed and became extinct in 90 seconds, and could not be rekindled by a temperature of 156°. The maximum of brilliancy obtains at a temperature of 99°, and is less intense and less permanent at temperatures superior to this grade. It may be inferred, that the temperature at which the animal dies is also that at which the luminous matter becomes particularly faint and evanescent.

When I plunged the luminous spherule in naptha, it was apparently immediately extinguished, but this seemed to be in consequence of the opacity of the fluid, for when withdrawn it shone with undiminished brightness. In solution of nitrate of silver, it shone for some minutes with undecreasing light; in etherial solution of gold, the luminous matter ceased to shine in 65 seconds—perhaps from a deposition of a film of the gold on the luminous spherule; in concentrated chromic acid, it remained luminous for several minutes—the acid being opaque, the duration was determined by withdrawing the luminous matter from the liquid. When I pressed the luminous matter into a flat disc, the light appeared momentarily, as so many minute scattered points, which became promptly evanescent from the rupture of the capsule including it.

Two spherulæ, with a portion of the interstitial matter, were detached from a glow-worm; on being withdrawn, they emitted light, but by a contractile power, soon became imbedded amid this substance, from which they could, however, be easily pressed, and then shone as at first. This luminous matter appears, from my experiments, to be of a spherical form, included in a capsule perfectly diaphanous, or transparent, which, when ruptured, unfolds the included luminous matter in a liquid form, of the consistency of cream. This luminous matter, there can be little doubt, remains *permanently luminous*, and the eclipse is entirely occasioned by the spherulæ in which the luminous principle resides, being withdrawn by a contractile movement into the darker recesses of the body of the insect, or being embossomed in the interstitial substance. In one instance, when the weather continued warm, the thermometer being above 70°, a glow-worm was accidentally crushed on the floor; yet though it perished, the light continued visible for three or four days. I remember to have picked up a lampyris noctiluca on the road from Ouchy to Lausanne. It had been destroyed by the foot, yet was nevertheless luminous.

I transferred a spherule into oil of olives in a small glass capsule, and viewed it under the lens of the microscope; it was opaque, of a bean shape and indented, slightly chromatic at the ends when illuminated by the reflector from below, being at one end fringed with blue, and at the other reddish yellow—the matter being itself yellow, and environed by minute air bells. The luminous matter being divided by a lancet, previously gilt with etherized gold, to prevent oxydation, in order to ascertain whether the hemispheres, thus separated, would discover a luminous annulus surrounding the central mass, discovered no such phenomenon, the luminous matter was instantly dissipated, and hence seems subtile and volatile, for a portion of the same being instantly transferred on a disc of thin steel into muriate of platinum, remained luminous for 85 seconds, while the rest had ceased to be so, in less than one-fourth of that period.

This last experiment proves, that though the luminous matter may, when undisturbed in its capsule, remain even for hours luminous, is volatile and subtile, when its containing membrane is lacerated or destroyed. The corrosive medium, even of a solution of nitromuriate of platinum, could act only on the exterior wall, and not affect the luminous matter imprisoned within its confines.

The yellow spherule, viewed by candle-light in oil of olives, seemed environed by a cloudy atmosphere. When, from extreme cold, it had ceased to be luminous in the dark, I held it immersed in olive oil in a glass capsule in my warm hand for five minutes, it then emitted a faint nebulous light. This resembles the character of some of those substances called solar phosphori; for instance the *chlorophane*, of which I possess a specimen that, held in the hand, evolves a fine green light.

We are wont to contemplate light as associated with the production of combustion, or the offspring of chemical action; but here is a light under a peculiar form, and independent of chemical conditions. The following appear to me corollaries deducible from the preceding detail of experiments:—

I. Light, as connected with the glow-worm,

is a subtile evanescent material principle, perhaps connected with a peculiar organized structure, or attached to a substance circumfused round the vitellula of the ovum, or integrating with it; unsupported by any chemical action, and confinable by the transparent film, or capsule, which imprisons it.

II. This light is permanent, and independent of any power possessed by the insect over it, except in so far as it can withdraw the luminous matter from the window, or transparent medium through which it is discerned, burying it in the interstitial matter, or secreting it under an opaque shell.

III. The light is not connected with any of the functions of animal life, as to its support or continuance, as with the spiracula or breathing apparatus, and even the extinction of life itself does not extinguish the power and property of emitting light.

IV. The luminous matter is not adherent exteriorly, but included in a capsule, which preserves it from extrinsic agency and contingency.

V. The light seems connected with peculiar organization, which elevated temperatures destroy, perhaps by decomposition, but which low temperatures only suspend temporarily. This very suspension, indeed, by cold, and restoration by warmth, and by a temperature equal to that of animal heat, goes far to prove a peculiar function, *inherent in the capsule*, and capable of educing and sustaining the phenomenon.

The use to which it is subservient in the animal economy, it is difficult to ascertain—"we see but in part." Its very existence, however, proves that it is a condition indispensable to its being. Providence has tipt the insect with living fire—a non-material ignition—burning, yet not consumed—even extinguished by a temperature which the animal system, with which it is so singularly interwoven, cannot withstand. It may be a "lamp to its path," to guide it to its food, subserving the additional purpose of warding off its enemies—while it may also be the luminous point that directs the nightingale to its proper prey.

LUMINOSITY OF THE SEA.

LUMINOSITY OF THE SEA.

———————

THE light of the sea has been ascribed to various causes—by some to phosphorescence, the effect of animal decomposition; to the imbibition of solar light, analogous to the diamond, and to an electric effect induced by friction; while others have more plausibly assigned it to the presence of luminous animals, and of these the cancer fulgens; medusa pellucens, hemispherica, &c. limulus noctilucus; salpæ, &c. have been described.

It is said, by Olof Wasserstrom, in the Transactions of the Swedish Academy for 1798, that the phosphorescence of the sea, in northern countries, may sometimes be owing to the small and very thin needles of ice which almost cover its surface, being broken in pieces by the agitation of the waves, and thus emitting a light, may assist in giving a luminous character to the sea! This is hypothesis strained to its utmost. The obtainment of marine animalculæ in high latitudes, is quite extraordinary. Capt. Scoresby, an accurate observer, gives us a curious calculation of their numerical array, almost beyond the lore of arithmetic to sum up their tribes. The language of the sacred text, in reference to their myriads, and the luminosity of the sea produced by the transit of a moving body through the volume of waters, is emphatically descriptive. " One would think the deep to be hoary:" " He," (referring to the ' leviathan,') " maketh his path to shine." The cause that produces the phenomenon in lower latitudes and warmer climes, must be operative in higher regions. Some fish, such as the herring, mackerel, whiting, &c. yield light in the incipient stage of decay. In this case, I am of opinion that it proceeds from adhering, perhaps parasitic, luminous animalculæ, the evolution of light being the effect of the slight increment of temperature produced by the commencement of animal decomposition. Dr. Ure has given us some curious and interesting remarks on this subject. " A solution of one part of sulphate of magnesia in eight of water, is the most convenient menstruum for extracting, retaining, and increasing the brilliancy of this light. Sulphate and muriate of soda also answer in a proper state of dilution with water. When any of the saline solutions is too concentrated, the light disappears, but instantly bursts forth again from absolute darkness by dilution with water. I have frequently made this curious experiment with the light procured from whiting. Common water, lime water, fermented liquors, acids even very dilute, alkaline leys, and many other bodies, permanently extinguish this spontaneous light. Boiling water destroys it, but congelation merely suspends its exhibition, for it re-appears on liquifaction. A gentle heat increases the vividness of the phenomenon, but lessens its duration." * These

* Dr. Ure's " Dictionary of Chemistry," article *Light*.

phenomena are by no means incompatible with
the idea suggested, as to their dependence on
luminous matter, connected with luminous
marine animalculæ, perhaps parasitic. We
know, for instance, in reference to the circumstance of *dilution*, that when the water which
contains the *gordius* evaporates, it shrivels and
dries up; it may then be preserved for any
length of time, but when transferred to a glass
of water, it soon lives and moves. A similar
thing happens to the *rotifer*. As to the rela-
tions of *heat* and *cold* to the luminous matter,
analogies will be found in my observations and
experiments on the light and luminous matter
of the glow-worm.

A writer, in No. 16 of the Scientific Gazette,
has the following remarks on the oyster:—
" An attentive observer recently remarked, on
opening an oyster, a shining matter, a bluish
light, resembling a star about the centre of the
shell, which appeared to proceed from a small
quantity of real phosphorus. On being taken
from the animal, it extended nearly to half an
inch in length; and when immersed in water,

seemed in every respect the same as phosphorus obtained from bones, &c. The oyster itself was perfectly alive and fresh, consequently the light could not proceed from any decomposition of the shell or animal, but must have resulted from some other source. On submitting this apparent phosphorus to a high magnifier, it was found to consist of three different animalculæ, one of which had no less than 48 legs attached to a slender body, a black spot on the head, which was evidently its only eye, the back exactly resembling that of an eel when deprived of its outer coating. The second insect, polypheme, had also a solitary eye and numerous feet—a nose resembling that of a dog, and a body made up of several rings. The third was very different, having a speckled body, a head resembling a foal's, with a tuft of hair on both sides. Each of these extraordinary insects was beautifully luminous, and altogether resembled a bluish star." I have myself repeatedly found luminous marine insects entangled among the minute *algæ* and *fusci* which invested the shell of the rock oyster. In the case cited, they

were doubtless introduced in the act of separating the shells. I have not unoften found
moluscæ that had thus entered the shell.

In the Transactions of the Wernerian Society, (Vol. IV. p. 171, &c.) Capt. Wauchope,
R. N., has introduced some remarks on the
phosphorescence of the sea, which go to prove
it connected with luminous animals:—" In September, 1816," says he, "in lat. 4º 52′ S., long.
9º 19′ E., I observed this shining appearance
very strongly, which induced me to draw a
bucket of water for the purpose of examining
it. I had it suspended, so as to have as little
motion as possible; when this was the case, it
shone very little; but the moment it was disturbed it shone with great beauty. I next got
a little lime-juice, and put a wine glass-full of
this acid into the bucket, when the shining
particles began to move about in all directions:
sometimes going only as far as the middle of
the bucket, then turning and taking a zig-zag
direction. These motions certainly had every
appearance of the depending upon the will of
an animal: they shone with much splendour,

and some appeared as large as the tip of one's finger. Another glass of lime-juice instantly destroyed them; for, at the instant the second quantity was poured in, the water appeared to be one blaze of fire, and no motion or disturbance could make it shine after this.

" I then drew up some more water, which shone as before; part of this I kept during the night in an open vessel, and part tightly corked up in a bottle; and the next night, on examining these two portions, I found that the water in the open vessel shone pretty brightly, but not so bright as it did; and that which had been corked up did not shine in the least—the want of air seeming to have killed the animals. They appear to me to be coated with some phosphorescent matter; for one of them I happened to rub upon my fore-finger, which left a streak of light, for a few seconds, as long as the first joint of my finger." The experiment with the acid, corroborates Dr. Ure's interesting one of a similar description, and seems to add plausibility at least to my view of the phosphorescence of the whiting, &c. Mr. Bywater, of

Liverpool, told me he had poured *acid* and *oil* on a portion of sea water in the docks, with a similar exhibition in the former case.

On my voyage from Leghorn to Civita Vecchia, I remarked that the Mediterranean was particularly refulgent, prior to a storm which we subsequently encountered; and this circumstance led me to investigate the phenomenon on our own shores. I was struck with the luminosity which the sea presented some years ago. It was succeeded by a gale; and may perhaps be considered its presage.* The sea sparkled with great brilliancy, and seemed to reflect the celestial scene from its bosom. A

---

* The remarkably luminous phenomena of the sea at Hastings, in December 1822, was succeeded by a terrific gale. I am in possession of numerous illustrative facts which verify the interesting conclusion, particularly one by a Gentleman of Macclesfield, who was much struck with the extraordinary luminosity of the sea, off the mouth of the *Mersey*, before a dreadful storm, in which two packets were wrecked on that part of the coast.

I believe *I may claim for myself* the *priority* of considering the luminous appearance of the sea, as to its increase of brilliancy or appearance on the coast, as connected with this new meteorological feature—the coming storm.

more attentive survey appeared to present at
least two distinct phenomena of this description.
One seemed to scintillate and was minute;
while the other exhibited an undulatory movement of the phosphoric kind, apparently com-
mencing at the centre, and diverging in concentric circles to the edge of the discs, which
seemed sometimes an inch in diameter. Immediately before the gale, I saw a solitary
gleam; but during its continuance could discover none, as to the waves that washed the
shore.

In reference to the connection of the luminosity of the sea with the storm, I may state that
I sustained this idea in my paper, transmitted
30th November, 1819, to the Wernerian Society, and since published in their Transactions.
About a twelvemonth or more ago, Dr. M'Cul-
loch promulgated the same opinion. The
following extract, from " Prince Maximillian's
Travels in Brazil," &c. would countenance the
same view of it:—" During a storm the sea
was very luminous, the intermediate surface of
the ocean seemed to be on fire. Day after day

the storm continued to rage with unabated fury."

I am inclined to attribute the increase of light in the sea, *prior* to the storm, to an *increment of temperature,* which I find the precursor of a gale.

On 12th July, 1823, at 6 p. m., 30 fathoms water, in St. George's channel, off the Point of Dromore; air, $55^0.5$ Fahrenheit; surface water 49^0; wind, S.W.

At 30 minutes past 8, p. m., off Mull of Galloway; 60 fathoms water; air 54^0; water $52^0.5$; wind S. S. W.,—commencement of gale; 13th July, at 20 minutes past 11, a. m., air 62^0, water 56^0,—off the Sand-banks of Liverpool.

From the edge of the pier at Stranraer, on the coast of Scotland, I took up a small portion of sea water, including some luminous substance. It was sometime, indeed, before I could recognize the existence of any foreign body, to which I could attribute the luminous effect; I however clearly perceived that there was no phosphorized oleaginous matter on the surface of the water. At length the shadow of an animal, in rapid movement, appeared depicted on the bottom of the basin; itself transparent as the medium in which it floated. It seemed to be a *beroe*, and identical with the species called *fulgens* by 'Mr. Macartney.—(Phil. Trans. for 1800, p. 260.) The animal, when at rest, exhibited a somewhat crescent form. The cilliary processes of the ribs, which, in swimming, described a tortuous motion, seemed to be those parts from whence the light was derived, but of which the whole body occasionally partook. The *beroe* died a few minutes after I had received it, which I attributed to the light of the candle rather than to increment of temperature in the medium. When taken up on the point of a probe, it had the appearance and nearly the consistency of jelly; it was diaphanous, and presented a spherical figure of about one-sixth of an inch in diameter.

The late Professor Smith, of Christiana, considered that the luminous appearance which diffuses itself over the whole surface of the sea, in the Atlantic, arose from a dissolved slimy matter; and that the most minute glittering particles, when highly magnified, had the appearance of solid spherical bodies.* I cannot doubt but these luminous particles originated with some of the *mollusca* and *crustacea* adverted to; and that they had been detached by the action of the waves, or friction, the consequence of other causes, as may appear in the sequel: another quantity of sea water presented to me the *medusa*,† perhaps the species called *scintillans :* I could discover no other kind. The medusa was about three-fourths of an inch in diameter. It died very shortly after I brought it home; and perhaps the light, as in the former case, was the cause. It is supposed that the *scolo-*

---

* " That luminous appearance which diffuses itself over the whole surface of the sea, arises from a dissolved slimy matter, which spreads its light like that from phosphorus. The most minute glittering particles, when highly magnified, had the appearance of small and solid spherical bodies."—*See Professor Smith's Journal, &c. p.* 258.

† On transfering a luminous medusa, taken off the mouth of the river Nen on the coast of Norfolk, to a glass of fresh water, the luminous matter in its descent contracted like a purse-mouth.

pendra electrica is destroyed by exposure to solar light; and Mr. Macartney observes that the medusa always retreated from the surface as soon as the moon rose; he also states that exposure to day-light deprived them of the power of shining.

By agitating the salt water containing the medusa, occasional gleams were exhibited. I also transferred it to a basin of *fresh* water, when it sunk to the bottom like a falling star. The effect here was of the most beautiful description. There appeared strings of minute beads of fire, like a chain illuminated by electricity. On stirring the fluid, these luminous points were disentangled, and displayed a pretty hemisphere of stars. The floating fires soon, however, ceased to lighten the fluid mass; agitation could not restore the effect, though a few drops of acid caused a solitary gem to twinkle. The medusa was transparent and gelatinous like the other. There can be no doubt but it was the tendrils, or feelers, which were vested with this brilliant and beautiful ornament: and I presume that these living fires originated here as in the cilliary processes of the beroe; and were subsequently, by some peculiar mechanical impulse, transfused over the transparent membranous sac or disc.

It seems quite certain, therefore, that the luminosity of the sea is a phenomenon dependent on the presence of luminous marine animals. The following interesting description, in further corroboration of the opinion, is extracted from the " Narrative of an Expedition," &c. by Captain Tuckey. London. 4to. 1818. p. 49:—" The *cancer fulgens* was conspicuous; in another species, when put into the microscope by candle-light, the luminous matter was observed to be in the brain, which, when the animal was at rest, resembled a most brilliant amythest, about the size of a large pin's head, and from which, when it moved, darted flashes of a brilliant silvery light. Beroes, beautiful holothurias, and various gelatinous animals, were also taken up in great numbers. Indeed, the Gulph of Guinea appears to be a most prolific region in this sort of animals."

We dare scarcely speculate, touching the design of this singular distinction. This much we know, that Almighty Goodness has made nothing superfluous, or in vain. The visitation of luminous animals seems connected with meteorological phenomena; and it would be interesting to ascertain, from different parts of the coast of Great Britain, what kind of luminous animals generally contribute to the effect.

ON

THE PHENOMENA

OF THE

CHAMELEON.

PHENOMENA OF THE CHAMELEON.

THE curious and interesting animal which is the subject of the present memoir, exhibits phenomena of a very remarkable kind. The earlier Naturalists could not overlook features so extraordinary, and accordingly it found a place in their pages. Fable has drawn on its resources, and poetry has consecrated its wonders by the witchery of song. There is not perhaps a more curious train of phenomena, nor a problem of more difficult solution. Its habits are very wonderful; its mechanism and structure of no ordinary kind; and the singular changes to which the creature is subject, involves a question of an uncommon complexion.

The following paragraph on the habits of the chameleon, I have extracted from the " Calcutta Journal," where it formed an article under the signature " H." " I have kept," says the author, " chameleons in a cage several months, narrowly watching them, and have placed them on different substances for the sake of experiment. I never saw an alteration in their colour, but merely a variation in their shade from a light yellowish green to a very dark olive green, the mottles were always visible, though changed similarly with the shades. The chameleon's tongue, which is nearly three parts the length of his body, is blunt at the end, and not unlike a common probe. From the end of it exudes a small quantity of matter, thick, clear, and glutinous. This he uses in obtaining his prey, which consists entirely of insects. He will remain sometimes for an hour with his tongue on the ground, and when a sufficient quantity of insects has settled upon it they are all drawn in and devoured. I have seen this animal dart at a fly settled upon a small piece of paper; the fly escaped, but the paper was drawn to the mouth by the cohesive liquid just referred to, and which I have several times particularly examined. The chameleon possesses the quality generally attributed to him—of a power of long fasting."

Some winters ago, I saw a living chameleon, for the first time, in Exeter 'Change menagerie, but it was sickly, and I could make no experiments upon it. In the month of July, 1824, I had this opportunity on a fine healthy chameleon, brought to Hull by the Captain of a ship from Sierra Leone, on the coast of Africa. It is not my intention to enter upon an anatomical description of its exterior appearance, it being sufficiently well known from the numerous specimens in the various cabinets of Europe as well as those in private collections. The eyes, in particular, are singularly constructed—each one being a rich and brilliant gem, set in a ring of gold, and enchased in a spherical socket; being adorned on the superficies of the capsule with radii uniting in this beautiful point of vision. Each ball performs its revolution entirely independent of its counterpart, and

when one eye shuts, the other still remains the
watchful sentinel. The globe, under such conditions, seems absorbed, and the convex superficies discovers a depression. Indeed, the eye
is remarkably intelligent and acute, and it
seemed to contemplate the ball of the thermometer, during the progress of the experiments,
with a curious interest. I have merely singled
out the organ of vision as not the least interesting feature of this elaborate structure, in
which such exquisite design is every where
manifest. Dr. Knox, see Vol. V. p. 104, &c.
" Memoirs of the Wernerian Natural History
Society," has discovered in the eye of this animal the *foramen centrale,* in the anatomical
dissection of a large specimen sent to the
museum of the University of Edinburgh by the
Marchioness of Hastings. Dr. Knox says,
" the whole of this very singular structure, *viz.*
the *foramen* and the fold of the retina, are
remarkably developed in the chameleon, and
actually much larger than in the human species.
There extends, also, from the entrance of the
optic nerve to the *foramen,* a fissure, which

however is not real, but apparent; this semblance of a fissure is caused by a remarkable thinness of the retina at this point. The retina around the *foramen* has adhering to it a quantity of black granulated matter, which it probably receives from the choroid.

" Exactly at the point corresponding to the *foramen*, the choroid is somewhat elevated internally, whilst it transmits externally a dark-coloured membranous canal, or tube, to the sclerotic. This is the only point at which I have found the choroid adhering to the sclerotic in this animal."

My experiments were made on the TEMPERATURE of the animal, as connected with the changes of colour depicted in such varying shades on the surface of the skin, and the magic of the necromancer's rod takes not the sense of vision more completely captive than do these ephemeral and sportive hues. I may premise, that I have in numerous experiments clearly and satisfactorily ascertained that each tint of the chromatic series of the prism even discovers a temperature peculiar to that colour, and in a constant ratio of progression. In the subsequent experiments, I employed a very delicate and sensible thermometer, which was instantaneously affected by the contact.

20th July, 1824, at 55 minutes past 4, to 10 minutes past 5, P. M.; temperature of air, $72^0.5$ Fahrenheit; ball in contact with either side, 73^0 to $73^0.5$ Fahrenheit; ditto, ditto, $73^0.5$ to 74^0 Fahrenheit.

As the tint varied from a yellowish-green to a deep pea-green, the side farthest removed from the source of light, *invariably*, in all my observations, discovered the *lightest* colour; and when that side was purposely turned toward the window, it assumed, in a short time, the darker shade, while the other softened down proportionally into the lighter tint; and when the light was equalized, the two sides were of similar complexion.

Air, 72^0; temperature of skin, 73^0 on lighter side, and $73^0.25$ on the dark-green side; yellow field of colour, $73^0.5$ Fahr. In another experiment, the yellow was $74^0.5$ Fahrenheit.

While moving on the floor it became very

opaque, and the still darker spotted bands exhibited a temperature of 75°.25. It is worthy
of remark, that where the ball of the thermometer rested, though the pressure was gentle
in the extreme, that spot became *snow-white*.

21st July, at 30 minutes past 10, A. M.; air,
69°.5 Fahr.; neck of the chameleon, 70°.5;
dark-coloured band, 71°.5; lighter parts, 71°.

In sun-shine, the bands, zebra-like, became
remarkably distinct, and the darker shades
indicated 74°, while the intermediate green
grounds oscillated between 72°.75 and 73°.5
Fahrenheit. I do not believe that the coloured
ground or floor traversed, at all affects the
evolved tint, excepting so far as the *light* reflected and modified from various coloured
surfaces, may operate differently on the circulation of the blood; because I presume that the
change of colour is in accord with the circulation of the blood as affected by the action of
light on the vital fluid through its membranous
envelope. As the circulation is languid, or
more active in its flux through the system, a
corresponding colour will be developed, to

announce this new phenomenon of chemical change, superinduced by the stimulus of light on the blood, and this tint will be a faithful index of its amount. The colour, too, I have considered to be the counterpart by which the temperature of the system is equalized. Hence, too, the negro tint* may subserve an equally important end in the system, by which the burning ray may be quenched.

In pursuing my remarks and investigations, I was surprized and gratified to find that my views and conclusions had been sustained by *Panarolus*, a Roman writer, quoted by Ogilby in his " Africa." The following extracts are drawn from the folio edition, published in 1670. Ogilby's work on Africa is a compilation from at least forty different authors, though he chiefly relies on Bellonius in his account of Egypt.

---

* Sir Everard Home has drawn some strange conclusions on this subject; by covering the back of his hand with a piece of *black* crape, and exposing it, thus covered, to the sunbeam, his hand was not scorched. Did it not occur to Sir E. to try *white* gauze or crape? Besides, what conclusions could reasonably be deduced from an envelope, extrinsic and external, thus applied?

" Chameleon is a Greek word, and signifies
' a *little lion.*' Bellonius says they frequent
about Cairo, and many other places, in the
hedges and bushes; it bears some little re-
semblance of the crocodile—from which it is
different in colour, head, tongue, eyes, and
feet; it creeps not, but walks upon all-four; the
head long and sharp, like a hog's; the neck
very short; and eyes, which, having no eye-lid,
can turn about on every side.

" This is a sluggish and dull animal, holding
the head carelessly, and the mouth always
gaping, lolling out the tongue, and so catching
flies, grasshoppers, caterpillars, palmer-worms,
and such like; instead of teeth, having one
entire jaw-bone, indented like a saw, but use-
less, swallowing whole whatever food it takes,
wanting both spleen and bladder, muting like
a hawk. The back hath a hard and rough
skin, beset with some few prickles; the two
fore-feet, Bellonius saith, have three claws inwards and two outwards; but the hinder feet
three outwards and two inwards, with hooked
nails, or talons. It hath a strange and ridicu-

lous manner of gait, or movement; for, stretch-
ing both feet on each side at once together,
and so alternately, the other makes a shuffling
gradation, one shoulder jetting foremost, the
other outstepping that, with a continual unto-
ward *hank* and *loose*, that it makes spectators
laugh, as if it were a match which side should
come first to the goal. But he is so nimble in
running up trees, that he seems rather to fly;
wherein he makes great use of his tail to lay
hold on boughs, especially in coming down;
whence we may gather, that the *chameleon* more
frequents trees than ground. Nor give the
motions of the eyes less cause of comical ad-
miration, for he does not as other creatures,
who turn both eyes at once after the same
object, but, something like our squinters, not
only looks two opposite ways at once, but more,
seeing right forward with one eye, and looking
up with the other aloft; another while to the
ground with one, and sideling with the other:
but, which is yet stranger, it will draw one eye
to its back, and make a survey behind, while
the other takes a prospect forwards. They

make at their meals also merriment, neither
picking as fowl, nor chewing like cattle, nor
sucking like lampreys and leeches; but with an
odd and sudden flutter of the tongue, shot out
near a hand's breadth, ingurges the caught
prey in a trice. This member being nothing
else but a hollow pipe, fleshy and spongy,
wherein are some sinews easier to shut together
than a gin or trap, because those nerves proceeding from the *os hyoides*, and running through
the cavity, draw the same after expansion back
again, with its prey sticking to a glutinous stuff,
wherewith it is covered. This refutes the
opinions of the ancients, who believed the *chameleon* lived by air; whereas, in truth, it lives
by such received nourishment as we have de-
clared.

" It appropriates to itself another peculiar
quality, in the opinion of some old writers, who
deliver that the chameleon changes colour
according to the several objects presented;
first in the eyes, then in the tail, after that in
the whole body. And this alteration of colours
many authors conjecture, and, among others,

the Roman *Panarolus* affirms to proceed from
the *systole* and *diastole* of the heart, which, according to sensibility of heat or cold, beats
quicker or slower, the quicker striking a red-
ness, whereas the slow reduces him to his own
natural ash-colour; for it retains that hue even
after death, though a little paler."

The new view of the phenomenon of the
chameleon, presented in the preceding pages,
might receive elucidation from many analogical
arguments. When the mind is surprized by
the tribute due to loveliness, and the circulation becomes thus affected, the conscious rose
instantly blossoms on the cheek of beauty.
The hectic flush on the cheek—the vermilion
lip—the pale ensign of the lily where the rose
once crimsoned—all concur to show that the
colours thus displayed on the exterior surface
are faithful indexes of the varied movements of
the circulating mass. The passions of the
mind affect the circulation of the blood, and
paint the visage. Thus, too, in inferior creation,
the buffalo and other quadrupeds are violently
excited by any thing *red*. When the male

turkey is irritated, and struts about in all the mimic pomp and pantomime of offended pride, the *caruncle* of the forehead relaxes, and this, with the naked parts of the head and neck, mount up from *blue* to an *intense red.*

Lord Byron has finely described the glow of youth and beauty, as affected by the passions of the soul :—

> " Her cheek all purple with the beam of youth,
> Mounting at times to a transparent glow,
> As if her veins ran lightning."

Tasso, too, admirably describes the contest of the rose and lily on the features of the beautiful Armida :—

> " Dolce color de rose ln quel bel volto,
> Fra l'ivorio si sparge, e si confonde
> Ma nella bocca, ond esce aura amorosa
> Sola rosseggia, e semplice la rosa."

Merrick, in Hudibrastic rhyme, thus adverts to the appearance of the chameleon by candle-light :—

> " I caught the animal last night,
> And viewed him o'er by candle-light,
> I mark'd it well—'twas *black as jet.*"

Now, though this may be considered merely as a *jet* of witticism, it should be mentioned, that Mr. Forbes, in his "Oriental Memoirs," adverts to the marked antipathy which the chameleon expresses to a *black* surface. "This," says he, "it carefully avoided, and when a black hat was presented to it, it shrunk to a skeleton and became as *black as jet ;*" an excitement in the animal analogous to *terror* in the human species, might well be deemed capable of producing such a change. The mere candle-light might not, after all, be the efficient cause, but rather some excitement entirely extraneous to it. Neither, on the other hand, would it be presuming too much to think, that this new illumination, to it unnatural and extraordinary, might operate remarkably on an animal susceptible of such exquisite sensibility in relation to the lights and shadows of the solar beam. The most singular effect of terror in the human species, as far as my limited knowledge goes, is that recorded in the " Journal de Medicine pour l'an 1817." It occurred at Paris, in the Hospital of the " Salpêtriere." A female, of advanced age, was so affected with horror, on hearing that her daughter, with two children in her arms, had precipitated herself out of a window, and was killed on the spot, that her skin, in a single night, from head to foot, became *as black as a negro.* This remarkable change continued permanent.

In that remarkable phenomenon, proceeding from a cardaic disorganization, wherein black and red blood intermingle, the skin, in consequence of a partial and incomplete circulation, is an intense *blue.* The fact, therefore, proves that the blood can develop and sustain a specific dye on the cutis.

I cannot conceive a finer illustration of my view of the case than the brilliant mockery of vision which is displayed in the dying dolphin, as life ebbs through the orifice of the bleeding wound, and is thus clearly shown to be dependent on the flux and reflux of the blood. Falconer has described the phenomenon in all the richest language of poetry.

I have lately been informed by an eyewitness, that the tints displayed in the dying dolphin increase remarkably in brilliancy and beauty, and with accumulating power to the terminal line of life.

The colour of the *iris* in the eye may resolve itself into a different question, but it is certainly connected with some characteristic feature of this strange microcosm. The fine *blue* eye, for instance, of the sprightly gazelle, introduced with such simple and beautiful pathos by Moore in his *Lalla Rookh*:—

> " I never loved a dear gazelle,
> To glad me with its dark blue eye,
> But when it came to know me well
> And love me—it was sure to die."

The views now submitted appear to me to be sustained by such forcible testimony that the evidence seems almost irresistible.

How diversified the interest and beauty which every where pervade the loveliness and grandeur of the creation of God! And what sublime pleasures are sacrificed by the non-observance of the wonders of creative omnipotence!

ASCENT OF THE SPIDER

INTO

THE ATMOSPHERE,

AND ON ITS

POWER OF PROPELLING THREADS.

K

ASCENT OF THE SPIDER, &c.

IF the views of the Entomologist reach no higher than the collection of a cabinet of butterflies, or the technicalities of mere nomenclature, we may consent to praise his diligence and laborious research, but must withhold our meed of approbation to the soundness of his views as a Naturalist. If entomology be confined within limits such as these, the study is dull, worthless, and contemptible.

Entomology has, however, a wider range and nobler field of usefulness; and, though some pseudo-philosophers have sneered at the *littleness* of the creatures thus contemplated, they show but their puny wit, and the feeble-

ness of their own understanding. Apart, they
may occasionally appear but " feeble folk;"
but, with a commission from Jehovah, become
" as the armies of the living God." The contemptible moschito may drive man to madness
—and the *terrible zimb* make the most savage
beast of the desert tremble and flee before it.
An army of locusts, denser than the storm-
cloud, and compassing an extent of many miles,
may lay waste, in a few short hours, the blossom
and promise of the year, and mercilessly con-
sign the myriad population of a vast empire to
pestilence and famine, and all the " bitterness
of death;" and, in its ravages, mock the devas-
tations of Turks, Saracens, and Tartars, and
claim mightier trophies and triumphs than an
Attala, an Alexander, or a Genseric. Nay,
more, the great features of the terrestrial globe
may be essentially changed by the humble
zoophite or polype, and man and all his proud-
est labours are mocked by the architecture of a
worm. Coral reefs and islands rise from the
ocean's unfathomable bed. These stony trees
soon accumulate into rocks and reefs, and

finally braving the surface of the deep, block up the entrance of harbours, and become destructive snares for the mariner: the waves of the sea rake up from their deepest abyss, sand and mud; and transporting sea-weed, finally to decompose, land their treasures at this new isle of the ocean; while the atmosphere scatters from its bosom the embryos of vegetation, which its friendly breezes had borne from distant shores, and it soon becomes an emerald isle, blooming with the refreshing verdure of luxuriant vegetation. Thus is reared a theatre for some future warrior, poet, or philosopher, to " strut his little hour upon," for man comes at length, and here plants the standard of his country, and claims it for his own.

The colour that vies with the brightest gem of the rainbow, we owe to an insect. The finest tissue, once balanced for its weight in gold, we receive from an insect—an island from another. What is sweeter than honey?—an insect prepares it. These are but a few of the benefits we receive from the insect tribe; but these exalt them in the scale of usefulness.

Nor are the subjects of entomology without their value in the arts that are mechanical. James Watt received his idea of the structure of the joints of the flexible pipes across the Clyde from that in the tail of the lobster—an animal which, though now removed to a new family, the *crustacea*, still maintains its former place in the system of some entomologists. We have, too, a peculiar spring constructed on that of the cicada, or grasshopper. More examples might easily be added, but these are enough.

By studying the natural history and economy of insects, we shall obtain more enlarged views of the beneficence and omnipotence of Heaven, and in their protection and preservation admire that Providence which watches over the minute as well as the vast and gigantic in the universe. It is not less delightful to contemplate the humble *lolium arenaria* guarding the confines of the sea by interlacing and intertwisting the sod that enamels its shore, as if commissioned by Providence to say to its world of waters, " hitherto, and no farther, and here shall thy proud waves be stayed,"—than to contemplate the " moon walking in brightness" amid a thousand twinkling worlds of light. If *size* were the touch-stone of excellence and the standard of appeal, then, as has been well observed, would " the horse be more excellent than his rider."

It seems to me that the phenomena of the ascent of the spider will be found ultimately connected with the meteorology of the atmosphere, and the observation of its curious habits lead to some useful practical results; and none dare pronounce on the ultimate importance and value of any fact — generally unobtrusive in the beginning, but often finally forming the mighty fulcrum of some noble power—like that of *latent caloric* in the hands of Watt.

If the physiology of insects be taken into the estimate, their singular structure and mechanism, their habits, their ingenuity, the mode of attack and defence in their curious wars, their amusements, and their cares and anxieties in providing for themselves and their dependents; a world of interest is unveiled. In this last most pleasing department, Messrs. Kirby and Spence have high claims on our grateful respect.

Among all the phenomena presented to the study of the entomologist, there are few to be found more interesting than those that are met with in the family of the spiders. Some of them are of a vast size—the Barbary spider is as large as a man's thumb. Haafner, in his Pedestrian Journey through Ceylon, thus describes the horned spider:—" Its brown rough body was more than six inches round, and its claws, the thickness of a quill, held a lizard, the flesh of which it was greedily devouring. I could plainly see its fiery eyes rolling in its head." I can remember that a group of the *aranea tarantula* once put me to flight, while botanizing at Portercole, a port on the coast of Italy, whither we were driven in a storm. The bite of this spider is said to be occasionally fatal, and its supposed cure by music has been often referred to. The Neapolitans have a dance called " Tarantella Napolitana." It requires violent exertion—the true cause of the expulsion of the morbific virus; being determined to the surface of the body, it is carried off through the medium of perspiration.

The *aranea avicularia* is covered over with hair, and catches small birds, which it devours. I remember to have once seen in Bullock's Museum, a spider of this kind with a trochilus or humming-bird entangled in its net.

The *aranea diadema* is found in gardens: it is sometimes met with of considerable size. A monstrous one of this species planted his net in the window of my room at Geneva; it was as large as a Spanish nut. I have held the aranea diadema under water for a considerable period; it not only seemed to suffer nothing by its submersion, but, on the contrary, set about deliberately weaving its web. "The jumping spider" is also a curious one of this singular tribe. The Barbary spider carries its young in a pouch or bag, like the opossum or kangaroo. The young spiders, after being nursed and nestled there, sometimes attack and destroy their parent. The female frequently destroys the male spider.* The detail which follows, is somewhat fabulous, and yet I know not whether it has ever been contradicted:—It is stated that the sexton of the church of St. Eustace, at Paris, was surprised at often discovering a certain lamp extinct in the morning. The oil appeared to be always regularly consumed. He sat up several nights, in order to discover the cause of its mysterious disappearance, and at last perceived a spider of enormous dimensions descend the chain or cord, and drink up all the oil. A spider of vast size was also seen in the year 1751, in the cathedral of Milan. It was observed to feed on the oil of the lamp: when killed, it weighed four pounds! and was

---

* The female spiders are, however, careful of their young. The sac, or bag, in which they deposit their eggs, is a tissue, impenetrable by water, spun by themselves, and this bag is their constant companion wherever they travel; some keep constant watch over their young. I remember to have witnessed this most interestingly exhibited in the case of a *Phalangium,* whose water-tight nest I rent asunder, and disturbed the included young. The female never stirred from the spot, but endeavoured to rally them round her, and restore tranquility and confidence.

afterwards sent to the Imperial Museum at Vienna.

The habits of the *aranea aquatica* are worthy of particular notice. This spider builds her residence at the bottom of the pond, and the web of which its tent is constructed is somewhat conical, like that of half an egg. From the atmosphere at the surface, the spider carries down, head foremost, a portion of air, and places it below the bell, and then another quantity, until it is thus inflated and filled with air, of course the air being thus deposited will ascend to the top of the bell, like that on the shelf of the pneumatic trough, and displace a proportionate volume of water, which will escape from the bottom or rather at the portal provided at the side for the ingress and egress of the spider: the air-tight bell is thus at length filled with air, and distended like a balloon; and, like the diver in the diving-bell, of which it is the principle, and might have, if observed and attended to, suggested its construction and proved its security; sits the spider, untouched by the water above and surrounding its aerial dwelling-place. From this abode *en ambuscade*, it pounces on aquatic insects, as the *caddis-worm*, &c., and devours them.

The *aranea geometrica* spins a very beautiful web, and it is indeed worth while contemplating its operations on a calm summer evening, in the snug corner of a window, a hedge, or paling, &c. An old Latin poet has said,—

" Nulla mihi manus est, pedibus tamen
Omnia fiunt,"—

and in allusion to the spider, the naturalist of the sacred volume has observed, " The spider taketh hold with her hands," &c.; and in these curious weaving operations, it is interesting to note how well the tibia and tarsus supply the place of hands. Radii diverge toward the periphery of several concentric circles, and these radii will be found to be more polished and glassy than those threads which intersect them, and are last woven. The spider having collected the threads emerging from the centre, retires to his hiding-place in one of the angles, and holding these threads by its tibia and tar-

sus, as a coachman managing his reins, is thus
informed by the vibration communicated, when-
ever an unfortunate and unwary insect is entangled. The particular thread of the web is
thus easily discriminated, and though the
spider may be so effectually concealed as to be
invisible to the victim, and the victim to it, it
will pounce instantly on it, glancing along this
single line. Though the web of the aranea geometrica is generally constructed of concentric
circles thus intersected, I have seen it somewhat in the form of a parallelogram, being a
complete tissue of threads woven in the form
of isoscles and scalene triangles.

That this curious tribe of insects may present
occasional phenomena connected with the meteorology of the atmosphere, my researches on
the *aranea aeronautica* would seem to prove,
and the manner in which some spiders carry
on their operations confirms the conclusion. If
the weather is likely to become rainy, windy, or
the like, the spider fixes the terminating threads
by which the entire web is suspended, unusually
short, and in this state awaits the impending

change. On the other hand, if these threads
are discovered to be long, we may conclude
that it will be in that ratio serene, and continue
so for about a week or more. If the spiders
be completely inactive, rain will likely follow,
but if, during the prevalence of rain, their
wonted activity is resumed, this rain may be
considered as of short duration, and to be
quickly followed by fair and constant weather.
It has been also observed that spiders regularly
make some alterations in their webs every
twenty-four hours, and I feel persuaded that
this is the case. If these changes are observed between 6 and 7 o'clock, P. M., they in-
dicate a clear and pleasant night. It is really
beautiful to observe, in a fine summer's day, the
threads that fan and flutter in the breeze from
the trees and hedges, and they are often
stretched across the road from hedge-row to
hedge-row, particularly in a morning. Spiders
have an ear attuned to the " concord of sweet
sounds." Hence, in the church or the concert
room, they are sure to let themselves fall from
the ceiling by lengthened threads; and if a

flageolet be played, spiders will hang motionless, having let themselves down not far from the spot where the instrument is played; in this way have those spiders perhaps been tamed that have become the only companions of the solitary in his cell or dungeon. Even the tight-drawn spider's thread, *may,* when struck by the passing wind, emit an Æolian note, audible to a spider's ear.

The ascent of the wingless spider into the atmosphere, is a fact unquestionable and un-questioned. It is one, however, as yet recorded without a solitary attempt toward its solution. I have consulted authorities in vain. Linnæus, Shaw, Donovan, and others, throw no light on the subject, nor indeed attempt a solution. The Edinburgh Reviewer, in his review of Messrs. Kirby and Spence's excellent work on Ento-mology, thus remarks:—" The flying of certain spiders, by means of their webs, is not the least extraordinary mode of motion possessed by spiders; nor, in truth, is it very intelligible, although the fact itself is unquestionable. In ordinary cases, the spiders spin their threads

slowly from organs adapted to that end, perforated with numerous holes, so that each
thread may consist of many thousand filaments.
The flying spiders, on the contrary, can dart
out the thread in a straight line for many
inches, in any direction, and then in some
unknown manner they follow it. In these cases,
when the animal and his chariot are wafted
away together by the winds, there is no difficulty. Our authors have thrown no additional
light on this difficult subject."

The gossamer-web was formerly believed to
be a tissue of " scorched dew;" hence Spencer,

" The fine net which oft we woven see
Of scorched dew."

Even Dr. Hooke said that the gossamer only
" much resembled a cobweb," and believed that
" the great white clouds that appear all the
summer-time might be of the same substance."

I have met with the aranea aeronautica, not
only in Great Britain and on the level of the
sea, but at Lausanne, in Switzerland. One
alighted near me in the steam packet, when

traversing the lake of Geneva, and several miles from the shore. I have also met with this curious insect on Montanvert, on the very verge of the *Mer de Glace.* Swammerdam and De Geer ridiculed the idea of the flight of spiders. Dr. Hulse first observed the property which particular spiders possess of propelling their threads into the air.

Dr. Martin Lister discovered that spiders were wafted aloft on this airy vehicle; and, in fine weather (in September) he found, more than once, a spider which, from its flight, he called " the bird." Afterwards he noticed that the insect, by elevating the anus, darted a thread from thence, and thus rose into the atmosphere.

From the highest point of the Cathedral of York, Dr. Lister beheld the gossamer-webs floating far above him.

Mr. White, of Selborne, confirms, by actual observation, Dr. Lister's account. He noticed a spider dart off from the page he was then perusing, and, though the atmosphere was tranquil, it rapidly ascended.

It has been considered that this property is not peculiar to one species, but that several spiders, when young, can so elevate themselves.

Mr. White conceived that spiders in their transit through the atmosphere could coil up their threads, and descend *ad libitum* from their aerial excursions, altering in this manner their specific gravity.

I am not aware that any have attempted to describe the gossamer-spider as a distinct and peculiar species, Bechstein and Starck excepted; but they seem to describe different species.

Thus, the former describes it as being the size of a small pin's head, having eight eyes disposed in a circle; body of a dark-brown colour, and light-yellow legs.

Starck describes it as extending more than two lines in length; eyes in the form of a square, two on each side, in contact with each other; thorax of a deep-brown colour, with paler streaks; the under side of the abdomen of a dull white, and a dark copper-brown colour above, having a dentated white spot running longitudinally down the middle.

Dr. Starck imprisoned several of these under a bell-glass, on a grass-plat, and he tells us they existed two months without food, though they took water greedily.

Mr. White observed a remarkable phenomenon on 21st September, 1741. Early in the morning the whole country was enveloped in a coat of cobweb, wet with dew. His dogs (being on a shooting excursion) were blinded by them. A delightful day succeeded. About nine o'clock A. M. fell *a shower of these webs,* (not single threads, but formed of flakes,) some nearly an inch broad, and five or six inches long; and such flakes continued to fall during the entire day. Baskets-full might have been collected from the hedges; and, from the velocity of their fall, it was evident that they were considerably heavier than the medium through which they descended.

The small spider with which these remarks are connected, has its eyes disposed in a *circle* somewhat elongated, •ᵕ•ᵕ• the body and legs, examined with a lens, are hairy, palpi bifid, and protuberant at the end; tarsus forked or clawed; legs, &c. somewhat translucent; abdomen and thorax glossy, and of a dark ferruginous colour; anal processes three; the femur and tibia have each two articulations.

Several of these spiders, included in a crowquill, were transmitted to Professor Jameson. Those called " Money Spiders" by the reapers in some parts of England, I presume to be the same insect.

The Rev. Mr. Kirby writes me, that he thinks the *aranea obstetrix* of Starck is that now spoken of: but the one described by Starck under this name is *striped,* and the eyes are arranged in the form of a *square,* which are sufficient distinctions. The subject of this article approximates more nearly to Bechstein's aranea obstetrix.

I shall take leave to call it " *aranea aeronautica,*" because, under the name aranea obstetrix the German naturalists describe *two different insects;* and I think it more than probable that Starck's aranea obstetrix is the *young* of the aranea geometrica, met with in hedges. The chief reason, however, for my proposing the assigned name, is the fact I have discovered, that its ascent and movement in the atmosphere are *essential to its very existence.*

I know well that the aranea geometrica does possess the power of propelling threads into the atmosphere, and of thus changing its locality or making its escape; and perhaps, too, the *young* of this insect may possess the power of taking an aerial excursion occasionally, but *if so,* I am persuaded it is a *very rare* event. The other is distinct and peculiar, and the numbers that occur in the atmosphere are such as sufficiently to account for the gossamer, and its beautiful and interesting phenomena. It cannot therefore be doubted, that those threads which glisten in the sun-beam, and float in the air from the hedges and hedge-rows, and the reticular tissue on grass, which, when sparkling with dew, refracts so beautifully the tints of the rainbow, are the work of the aeronautic spider.

As a proof that these wingless " birds" are more numerous than may be generally suspected, I may merely mention, that, in the

month of July, 1822, on the top of the coach
from Kidderminster to Stourbridge, (a distance
of only nine miles,) there fell on, or near me,
thirteen aeronautic spiders, all of which I
caught, and imprisoned in chip-boxes I carried with me for that purpose. This species
of spider may be frequently met with in *coach-
offices*, having alighted on the passengers, or on
their luggage.

Connected with this question, I may mention
a curious phenomenon that I witnessed on the
16th September, 1822, at Bewdley, Worces-
tershire. Between the hours of 11, A. M. and 2,
P. M., the whole atmosphere seemed to be a
tissue of cobwebs, which continued to fall in
great numbers, and in quick succession. The
temperature was 72° Fahrenheit. Some of
these were single, others branched filaments,
occasionally seen to extend from 40 to 50 feet
in length. Others were woolly films, or flocculi:
some fell slowly, and others more rapidly.
This was first noticed in the market-place, at
Bewdley; and, on repairing to the adjoining
fields, I found the same phenomenon, and my

clothes were most curiously invested with a net-work of spiders' threads.

In a communication to the Rev. J. J. Freeman, of Kidderminster, I remarked this circumstance; and the following is an extract from his letter to me, dated 18th September, 1822: " The fall of cobwebs was also observed here on Monday. A gentleman told me he was obliged to wipe his face several times while walking in his garden about 12 or 1 o'clock, such quantities continued to fall on him."

On the 19th of July, 1822, (the anniversary of the Royal Coronation,) the yeomanry, at 1 o'clock, P. M., were drawn up in the market-place at Kidderminster, and fired a *feu-de-joie* on the occasion. This had the effect of bringing great numbers of the aranea aeronautica from the aerial regions; very many I picked up from the pavement, when the yeomanry had withdrawn; and several took refuge on the table where I was engaged reading, near the window of the hotel, which was then partly open.

I have stated that a free and unrestrained privilege of ascent into the atmosphere is a condition essential to the very being of these remarkable insects.

The blaps mortisaga, it is known, will live *three years* shut up, and without food. I have kept the aranea diadema two months under similar circumstances. An entomologist informed me he had kept a spider three months without food; and indeed this insect has been preserved alive *upwards of a year* confined, and wanting nutriment.

The aranea aeronautica, however, I have invariably found, is impatient of confinement, and will *die*, whether imprisoned in a chip-box or glass-tube, (showing that *light* does not affect the question,) sometimes in *twenty hours*, or at most in *two or three days*.

I introduced one of the aeronautic spiders under water; and though it remained there upwards of a minute, it did not appear injured; and when withdrawn, soon let itself fall from a point, by means of a thread.

Posited gently on water, at 66⁰ Fahrenheit, it remained on its surface, without attempting to escape by the propulsion of a thread. It

took repeated springs forward, and then re-
ceded, patting the water rapidly with its tarsus,
in the manner of the squirrel.

In water at 67⁰ Fahrenheit, it was quiescent.
When reposing at the bottom of a tumbler of
water, there issued from between the palpi an
air-bell, which, expanding, carried the spider
to the surface; the aerial appendage thus dimi-
nishing the specific gravity of the aggregate,
and affording a striking elucidation of the habits
of the aranea aquatica.

An aeronautic spider being put into water at
94⁰ Fahrenheit, remained at the bottom of the
vessel, sometimes at rest, sometimes locomotive.
It projected a thread upward, and by that
means, *sailor-like*, wound itself, resting at intervals, to the surface of the water. At the
close of the experiment, the temperature of the
fluid had fallen to 86⁰ Fahrenheit.

One of these spiders, by candle-light, darted
instantaneously a thread to the ceiling of the
room (eight feet high); it described an angle
of about 80⁰ with the horizon. By means of
the combined act of the tibia and tarsus, " guid-

M

ing them wittingly," the thread was made to spin with great rapidity on its axis; and during this period it moved gradually toward the vertical plane, and, being thus highly twisted, formed a stronger medium of escape.

During my stay at Chester, while I was experimenting with an aeronautic spider, during a warm day, and brilliant sunshine, about noon; my room door was a-jar, and the insect, in the act of propelling its threads in all directions, suddenly darted one toward the door, in the direction of the influx-current, perfectly horizontal, and in length ten feet. The angle of vision being particularly favourable, I observed *an extraordinary aura or atmosphere round the thread*, which I cannot doubt was *electric.*

There are many phenomena that dispose us to believe the thread to be electrified. The following diagram will afford a representation of one of my experiments made in a room. The deflection from the vertical plane in the direction of the dotted diagonal line, is, I think, to be ascribed to the attraction obtaining be-

tween the wall as the conductor and the electrified thread. The aeronautic spider threw
out two threads toward the ceiling, one perpendicular, and the other inclined; and then
let itself fall from the end of a quill, resting at
a woolly ball, and from thence projecting a
horizontal thread, subsequently descended.

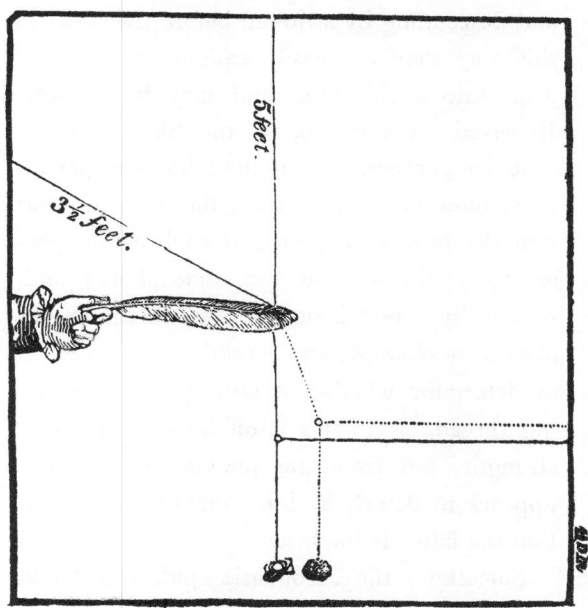

I first began my experiments and observations on this curious subject on the 2d of June, 1822. One of these spiders alighted on me, and glanced off from my hand with considerable rapidity: thermometer 77° in the shade.

It is impossible to walk in the fields without being saluted by several of these insects; they will be chiefly noticed by alighting on the hat, and descending by a thread before the face: in this way they are easily caught, as they will drop into a chip-box, and may be secured. Received on a pencil, or the like, they will soon be perceived to oscillate like the pendulum; oftentimes rising from the perpendicular into the horizontal plane, at each ascent projecting a thread into the atmosphere; and, finally, by a twitch or jerk, the insect breaks from its anchorage, and ascends. It is difficult to determine whether it *bites off* the connecting thread, or breaks it off by main physical strength; but from the sudden twitch which appears to detach it, I am inclined to believe that the latter is the fact.

Sometimes the aeronautic spiders will take

their flight immediately from the surface on which they alight, if the day be warm and sultry; but generally they descend to from 6 to 18 inches, perhaps the better to *insulate* themselves, and that, suspended by a pliant thread in free space, they may more freely propel their threads into the atmosphere.

Not unfrequently the propulsion of a solitary thread will bear them aloft, but the air must then be very warm, and its electric character be high. Sometimes the ascent is rapid, and cannot be followed by the eye; at other times it is slow and majestic. Occasionally the ascent is quite vertical, and at other times the animal sails on the bosom of the air, either in the horizontal plane, or at angles more or less acute.

It will be also found that there are *particular seasons* of the year best calculated for this singular exhibition: spring and autumn are these periods. In summer I have frequently found it utterly impracticable to determine their ascent. When they have detached themselves, after several vibrations, they have fallen to the ground like a dead weight. One day (in May,

1823,) this remarkable fact was determined in the case of nearly a dozen which I picked up—all that I experimented with on that day.

The insect seems to be aware when the threads are sufficient for its ascent : perhaps the temporary suspension in the horizontal plane may communicate the information.

The aeronautic spiders make their appearance early in the season. One fell on me in the beginning of March, 1823, while I was perambulating the streets of Bath.

I have shown the phenomena of the aeronautic spider to several persons, among others to Mr. T. Hopkins, of Kidderminster, and Mr. T. Brown, of Cirencester. To the latter, indeed, I one morning, in a very short period, pointed out the phenomena in five or six different instances.

Several circumstances concur to show the phenomenon of ascent to be *electric.* The propelled threads *do not* interfere with each other; they are divellent, and this divergence seemed to proceed from their being imbued with similar electricity; and the *character* of that electricity appeared to me to be an interesting subject for subsequent research.

The aeronautic spider, brought near a candle, descends by its vertical thread, winds it up, and descends again very rapidly and repeatedly. The flame seemed not to attract a short upright thread, though the *finger* brought near, did. Placed nearer to the flame, the animal seemed incapable of descending farther, but moved circularly.

The point of a gold wire was brought near to the vertical thread, in one experiment, above the spider in the act of escape to the ceiling of the room. It evidently disconcerted its progress, and the animal seemed agitated and unable to ascend. On removing the point the insect soon made its escape.

When the conducting wire is brought near to the thread by which it suspends itself, but, above all, to the flocculi or balls, the thread is considerably deflected from the perpendicular, and the horizontal fibre is attracted by the point.

When a metallic conductor is brought near

to the suspended spider, it disarranges its projectiles, and the insect, conscious of some counteracting agency, promptly coils up its threads.

When a stick of excited sealing-wax is brought near ,the thread of suspension, it is evidently *repelled;* consequently the electricity of the thread is of a *negative* character. The descent of the thread is instantly determined by bringing over it the excited sealing-wax; and if strongly excited, and the spider let fall on its surface, it bounds from it with considerable energy.

On the 3d of July, 1822, at 4, P. M., thermometer 66° Fahrenheit, two aeronautic spiders, on separate threads, were brought near to each other; a mutual repulsion supervened; and when one was brought in momentary contact with the other, it immediately fell lower in the perpendicular plane.

An excited *glass*-tube brought near, seemed to *attract* the thread, and with it the aeronautic spider. When the insect was thus *positively* electrified, the rapidity which marked its descent, and extent of thread spun out, which I frequently coiled up, was truly astonishing, being at least 30 feet in length.

In one experiment made, the ascent of the insect was so slow and tranquil, from the humidity of the lower atmosphere and wetness of the terrestrial surface, that I could easily catch it by following its progress: it moved in a plane parallel to the point of departure.

On the 4th August, 1822, at 3, P. M., thermometer 66^0, the ascent was slow and beautiful, the little aeronaut rising regularly in the vertical plane. It was distinctly perceived from the steady fixation of the eye, and favourable angle of vision, until it had attained an elevation of 30 feet at least, and was finally lost in the vanishing point of elevation.

A variety of other phenomena unite their testimony in favour of the conclusions I have formed, and from what I consider the direct method of induction.

Were the thread not electrical, I do not see how it could be propelled through the atmosphere in the vertical plane, and remain there, contrary to the laws of gravitation. It is indeed

remarkable, that the threads should remain in the precise plane in which they are propelled, nor ever swerve from that path. The constant relative separation finds an analogy in similarly electrified pith-balls, or the divergence of the filaments in the case of a glass plume, &c., placed on the conductor of an excited electrical machine. The undulations of an agitated atmosphere disturb rather than favour the ascent of the little aeronaut. The electric or calorific state of the atmosphere will be found always to modify the phenomena. The transit of the thread through a resisting medium, without its suffering deflection in its path, proves it to be imbued with a power superior to, and able to combat with or overcome, that resistance.*
The instantaneousness of the propulsion of the projectile can find no reasonable cause but in the subtilty of electricity and excitation of that

---

* The friction sustained in its sudden propulsion through the resisting atmosphere would alone be capable of investing it with electricity. I have seen, in the case of a fibre of very fine-spun glass, suddenly drawn upward, that it continued *vertical,* and I found it *electrical.*

mighty power. An illustration somewhat connected with the question is found in the propulsion of fine threads from melted wax connected with the conductor of the electrical machine in action, the threads being received on the surface of paper.

When the superior regions of the atmosphere are charged with *positive* electricity, while the threads are imbued with *negative* electricity, ascent into the atmosphere becomes a *necessary consequence.*

It is difficult to ascertain what part this ascent into the atmosphere subserves in the economy of nature. The spider may mount on high in its electric chariot to thin the ranks of those clouds of the genus *staphylinus,* which if suffered to remain undisturbed, *might, for ought we know,* increase even to the destruction of us and ours. Whether the spider becomes food for other insects, remains a question; but I have, in *several instances,* found that, while it oscillated, some prowling *vespa,* or wandering *musca,* has snapt up and carried off my " ARANEA AËRONAUTICA."

ON TORPIDITY,

AS CONNECTED WITH

THE "TESTUDO GRÆCA,"

OR

COMMON TORTOISE,

&c. &c.

~~~~~~~~~~~~~~~~~~~~~~~~"Sic sine vitâ,
Vivere quàm suave est, sic sine morte mori."

N

ON TORPIDITY, &c.

THERE are few topics less understood than the torpidity of animals, though it is one of the most curious features in their physiology. This strangely protracted sleep is a phenomenon remarkable in itself, but remains as yet unsuccessfully questioned.

As the following observations on the peculiar habits of the tortoise, and experiments made on the temperature of the dormouse, animals that become torpid, may eventually aid the question, so some previous remarks on the phenomena of torpidity may be in proper place.

In the insect world we are presented with numerous instances of torpidity, which ought, however, to be carefully discriminated from *hybernation.* Toward the close of autumn, when winter is pressing on its confines, insects are busied in looking out for safe retreats, wherein they may securely pass the brumal period of the year. Some of these, thus sheltered from the storm and surrounded by non-conducting media, enjoy a temperature unhinged by meteorological contingency. Others, however, become torpid, lose all sense and motion, and repose in this death-like state during the months of winter. Some insects pass the winter in their perfect state, as coccinellidæ and curculionidæ, &c. Others, again, do not. Several of the coccinellæ are solitary during summer, but are found in winter aggregated. Some of them are discovered out of winter quarters, even after the severe weather has set in. At this moment I have a coccinella duo-punctata, which was caught wandering in my bed-room, and it is now before me, active and lively, enjoying itself under a glass among the verdure of some cress and mustard, raised artificially on wetted flannel. The weather, too, has been unusually severe, as the following statement of the thermometer and hygrometer will show:—

Paisley, January 1826.

				Ther.	Hyg.	
Jan. 14,	5	A. M.	....	6°	...	In an adjoining field.
...	...	10	A. M.	....25°	...70°	There had been a good fire in the room for two hours.
...	...	3	P. M.	....33° 50′	...60°	
...	...	12	P. M.	....17° 50′	...85°	
...	15,	8¼	A. M.	....15°	...90°	When the window-shutters were first opened.
...	...	10	A. M.	....25°	...73°	

... ... 4 P. M. In an adjoining field, appended to a stake, one foot from the ground, the thermometer stood at 20°; horizontal on the ground, 21°; the ball buried in the hoar frost, 22°. Dense fog.

...	...	6	P. M.	....25°	...84°	
...	...	10	P. M.	....22°	...89°	
...	16,	10	A. M.	....26°	...70°	
...	...	4	P. M.	....38°	...65°	
...	...	12	P. M.	....36° 50′	...85°	Thaw.

Note.—The hygrometer is that of De Saussure.

The following remarks on the phenomenon of insect torpidity are by a most diligent and careful observer.* " The first cold weather,

---

* Mr. Spence. See " An Introduction to Entomology," &c. 8vo. London. 1817. Page 442, Vol. II.

after insects have entered their winter quarters,
produces effects upon them similar to those
which occur in the dormouse, hedge-hog, and
others of the larger animals subject to torpor.
At first a partial benumbment takes place; but
the insect, if touched, is still capable of moving
its organs.*   But as the cold increases, all the
animal functions cease. The insect breathes
no longer, and has no need of a supply of air;
its nutritive secretions cease, and no more is
required; the muscles lose their irritability;
and it has all the external symptoms of death.
In this state it continues during the existence
of great cold, but the degree of its torpidity
varies with the temperature of the atmosphere.
The recurrence of a mild day, such as we
sometimes have in winter, infuses a partial
animation into the stiffened animal: if disturbed,
its limbs and antennæ resume their power of
extension, and even the faculty of spirting out
their defensive fluid is re-acquired by many

---

* In the case of my tame dormouse, when semi-torpid, its
distended limbs contracted slowly to their former position.

beetles. But, however mild the atmosphere in winter, the great bulk of hybernating insects, as if conscious of the deceptive nature of their pleasurable feelings, and that no food could then be procured, never quit their quarters, but quietly wait for a renewal of their insensibility by a fresh accession of cold." This description seems exceedingly accurate and just.

It has been supposed that bees become *torpid* in winter. This is certainly a mistake, and the very phenomenon of their living in such perfectly constituted societies goes far to render the opinion questionable. Their very treasury of honey is a proof of the same kind. What can they provide against but the winter of the year, when not a flower unfolds and not a sweet is disclosed ? " She therefore provideth her meat in the summer, and gathereth her food in the harvest." I know to my cost, that if the hive be not defended against cold, or be stocked insufficiently to support a genial temperature, it will inevitably perish, and though I have restored to the hive, bees that have been frozen, this is only accidental. According to Huber, the warmth of a well-tenanted hive in winter is 86⁰ Fahrenheit, and my own observations confirm his statement. John Hunter found that a hive lost, from 10th Nov. to 9th February, more than 4 lbs. in weight. Ants become torpid, according to the same author, at about 27⁰ Fahrenheit, but by congregating together endeavour to preserve themselves from the severity of cold.

The eggs of insects are remarkably provided for and defended against cold. Spallanzani exposed the ova of the silk-worm, with those of other insects, to the action of a freezing mixture, 38⁰ below zero, Fahrenheit, but they were neither frozen nor had their fertility impaired. Insects may be frozen into solid masses of ice, and yet revive. In Capt. Franklin's Narrative, we see this extraordinary fact in the instance of a congealed icy mass of mosquitoes. I dug out from the *Mer de Glace*, last summer, a species of musca, which, by careful application of temperature, I have no doubt would have survived. The phenomena of life altogether is a constant succession of miracles. The sphex ichneumon punctures the body of the caterpillar and deposits its ovum, and the young larva devours the interior of the caterpillar, leaving only a husk, case, or skin; yet, though all the interior parts are thus devoured, the caterpillar will walk and eat. Snails in their shells have been thrown into a drawer, and having thus reposed for fifteen years, have recovered on immersion in a basin of water. When the water containing the gordius, or horse-hair eel, evaporates, it shrivels and dries up, and may be thus kept for any length of time, but when put into a glass of water it is soon as lively as ever. I found the gordius in the lake on the summit of the Great St. Bernard. The monks at the hospice have observed a cold there of 20^0 minus zero, (it was 18^0 last winter,) or 52^0 below freezing, yet this animal must have slept under this icy chain. The *rotifer* shrivels up as the water evaporates, and becomes like a bit of dried parchment, but when exposed to moisture it lives again—the extraordinary transition has been repeated eleven times. Lister states that he has found caterpillars so frozen, that when dropped into a glass they clinked like stones, and yet nevertheless revived. Mr. Stickney, of Holderness, in Yorkshire, who is a very intelligent and cautious observer, as I can state from personal acquaintance, exposed some of the larva of the *tipula oleracea* to a severe frost, which congealed them into ice. When fractured, their whole interior was found to be frozen—yet several of these were re-animated. Bonnet exposed the pupæ of *papilio brassicæ* to a cold amounting to zero of Fahrenheit—they became masses of ice, yet produced butterflies.

Spallanzani has made the curious observation, that insects re-appear in spring at a temperature considerably lower than that at which they retired in autumn. Agreeable to what has been mentioned, it is to be regretted that authors have not discriminated " between the *state* in which animals pass the winter, and their *selection* of a *situation* in which they may become subject to that state."

There can be no doubt that a continued
artificial warmth would prevent many animals
from entering into a state of torpidity; and the
insect, so destructive to the favourite of our
flowers, the *aphis rosæ,* if exposed in winter to
the inclemency of the season becomes torpid,
while in green-houses, &c. it preserves an active
and animated being. It cannot be owing to
any effect of cold, &c. *previously felt* that they
are induced to make preparation for their long
repose: it must be by a *law* of their being,
imposed by Providence, " that *previously* to
becoming torpid they select or fabricate com-
modious retreats precisely adapted to the con-
stitution and wants of different species, in
which they quietly wait the accession of torpidity, and pass the winter." It has been very
properly remarked, that the fact of insects
hybernating at the close of autumn, and their
not doing so in summer, when intense cold and
frosty nights prevail occasionally, is a sufficient
refutation that *cold* alone cannot be the exclusive agent. " We may say, and truly,
that the sensation of fatigue causes man to lie

down and sleep; but we should laugh at any one who contended that this sensation forced him first to make a four-post bedstead to repose upon."*

" A continuance of life," says Dr. Reeve, " under the appearance of death, a loss of sensibility and of voluntary motion, or suspension of those functions most essential to the preservation of the animal economy—these are the phenomena which accompany the torpid state, and they constitute one of the most singular problems in the whole range of natural philosophy."† When we contemplate animals shut up in their subterrene abodes during the brumal period of the year, we shall find that in those which become torpid the temperature of the skin is reduced to a low ebb, the circulation of the blood is entirely suspended, and respiration at an end; the torpid animal has contracted its limbs into the form of a ball, and

---

* *Ubi supra*, page 462.

† An Essay on the Torpidity of Animals. 8vo. London. 1809. Page 6.

is generally rolled up in some substance dry and non-conducting as to temperature. These nests are formed of dried leaves, grass, &c. The dormouse I found always rolled up during the season of repose—this is by day, and it had carefully wrapped the dry moss round itself. Perhaps one of the most extraordinary phenomena of this description is exhibited in the case of the dipus canadensis, or the canada jumping mouse, which has been found twenty inches beneath the surface of the ground, completely enveloped in a ball of clay nearly an inch thick. Now this material could not permit the functions of respiration to go on, though a loose and spongy texture, as of a sandy soil, might be supposed to minister, although slowly, to this process. Of the same description precisely, is the repose of the toad, lizard, &c. in the solid rock, some hundreds of feet below the surface of the ground. In such cases, it seems clear, respiration must have long ceased, for no sooner is their prison-house unsealed, than that which was first discovered an apparently lifeless and motionless form, begins to

o

move, and at length exerts its functions, which seemed as hermetically sealed as its lapideous dwelling-place. It is quite clear, also, that torpidity is preserved inviolate in this enclosure by the uniformity and unchanging temperature in which it is found—from a two-fold cause, the nature of the lapideous mass itself, and the depth below the surface where it reposes. Were such rocks formed by *fire*, as the Huttonian would endeavour to persuade us is the case!—it is quite clear that the animal must have been sealed up contemporaneous with the formation of the rock in which it is imbedded. A toad thus circumstanced was found *under the coal seam*, in the iron-stone over which it rested, in a coal mine at Ayr (at Auchencruive.) What then becomes of the speculation that coal was formed, as the Huttonian geologist tells us it was? Increase of temperature, and contact of an atmosphere acting with all its electric and hygrometric and barometric vicissitudes, seem essentials to the revival from torpidity. The toad and the lizard are the animals found sealed up in rocks, and this

question is one of a most interesting kind. It
is quite sufficient of itself to unsettle and unhinge many of those wild speculations touching
organic remains, which have been precipitately
and incautiously assumed in geological science.
They seem occasionally unsealed from the
rock, to tell us that they were as indigenous to
the soil, in a former and far distant age, as they
now are, and that they were the tenants of the
soil when the rock was formed. Similar attes-
tations come from the vegetable kingdom. A
gentleman in the Isle of Man, with whom I
have the pleasure to be personally acquainted,
dug up, from a depth of 16 feet, in a solid peat
moss, some grains resembling seed, that were
found imbedded in the most solid portion of it.
He sowed them in a garden-pot, and they grew
up into the *rag-weed*, and who can tell how
many centuries they had thus been hermetically sealed? The functions of germination
and vegetation were suspended, not destroyed.
The torpidity of toads and lizards for ages in
the solid material of the rock, and the repose
of seeds under ground for so extended a period,

are merely an elongation and extension of the same change and phenomena as are displayed at regular intervals every year, in the sleep of the tortoise, hedge-hog, dormouse, marmot, and others, or in the insect and reptile world, or in the repose of vegetation in the bud.

The *erinaceous europeus*, or hedge-hog, wraps itself most curiously in a ball of hay, with which it is supplied; the intertexture of the fibres is remarkably close, and the exterior surface quite smooth. I had once a tame hedge-hog, and its habits a good deal resembled those of my dormice. It took its food always by night.

Dr. Reeve, in his Essay on the Torpidity of Animals, has classed the phenomena of torpidity under the following heads:—

I. The temperature of hybernating animals is diminished.

II. The circulation of the blood becomes slower.

III. The respiration is less frequent, and sometimes entirely suspended.

IV. The action of the stomach and digestive organs is suspended.

V. The irritability and sensibility of the muscular and nervous powers are diminished and suspended.

On each of these a few remarks will be made, to preserve a somewhat arranged form. The causes of torpidity may be glanced at, and general deductions made.

I. Mr. Hunter ascertained that the temperature of a hedge-hog, at the diaphragm, in summer, was 97^0 Fahrenheit, when the thermometer in the shade was 78^0; and the air being at 44^0, the animal became torpid, and its temperature was $48^0.5$ Fahrenheit; the air being 26^0, the temperature of the animal was 30^0 Fahrenheit. Pallas, in the case of a torpid hedge-hog, found the skin under the belly $39^0.5$, and Spallanzani limits the decrease of temperature during torpidity to 36^0 Fahrenheit. I have stated the temperature of my dormice at 102^0 to 104^0 Fahrenheit; Dr. Reeve found it 101^0. It will be perceived, that the animal, in a semi-torpid state, was at one time $62^0.5$, and at another 69^0 Fahrenheit. Dr. Reeve found the skin of the dormouse, when rolled up and torpid in winter, 43⁰, 39⁰, and even 35⁰, on the exterior surface, but in the stomach 67⁰ and 73⁰—an approximation to my results: marmots have a temperature amounting to 101⁰ and 102⁰, and this sinks in the torpid state even as low as 43⁰. It would appear, that though the exterior surface of the animal be of a similar temperature with the ambient medium, the internal temperature is higher. Spallanzani found the wood-mouse, (which in Italy becomes torpid in November,) in a torpid state, when the air in its cage was 43⁰, and the temperature of the belly was 45⁰.

It seems clearly demonstrated, that in torpidity the temperature of the body is reduced.

II. In the case of the *mus cricetus*, when active and irritated, its heart beats 150 strokes in a minute; while in a torpid state the number of pulsations are reduced to 15. In the *bat*, the number of beats may be stated at 100, but when torpid these amount to only about 14. Dormice have a very rapid pulse; but when semi-torpid, the number of beats may be considered about 31. The action of the heart ultimately becomes imperceptible, and before this period the pulsations are limited to about 16. It is not probable, that though the circulation be reduced to its lowest pulse, and the diastole and systole become inappreciable, that the blood ever congeals, though it may be more dense. Spallanzani informs us, that if the blood of the marmot be subjected to a temperature even higher than that of the lungs of the animal, it is frozen, but that it is never congealed in their torpid state, even after exposure to a cold several degrees below the zero of Fahrenheit.

These facts show plainly, that though the circulation be, in a state of torpidity, reduced to a minimum quantity—that agency, *superadded* to their material organization, which we call *the principle of life*, preserves the circulating fluid in a liquid form.

III. By the experiments of Spallanzani, torpid bats lived seven minutes in the exhausted receiver of an air-pump, in which another bat perished in less than one-half this period. Some animals will support such an attenuated medium for a long period: I found this in my experiments with the *grasshopper :* the respiration seemed hurried and laborious, but it was not injured after a long period of imprisonment. Spallanzani confined torpid marmots for four hours in media of carbonic acid gas and hydrogen without injury, while when wakened in the gases they perished. The same philosopher found no endiometric change on four to five cubic inches of atmospheric air, after torpid bats and marmots had remained in it three hours. In 1795, Spallanzani found that torpid dormice, after being exposed to a temperature below freezing, remained uninjured in media of carbonic acid gas and azote over mercury; the internal surface of the glass was not dimmed by any moisture, consequently breathing must have been suspended. Sir J. E. Smith, President of the Linnean Society, has found that the respiration of a tortoise in winter was always slower than in summer.

It may therefore be fairly concluded, that in torpidity respiration ceases.

IV. Mr. Hunter introduced food into the

stomachs of lizards before torpidity was de-
clared, and on being subsequently examined
during this period, it was found unaltered. In
the case of the dormouse and marmot, &c.
no food is taken immediately before torpidity
commences, and at the close of torpidity, in
spring, the stomach, &c. are found empty. I
have cited a similar fact in respect to the tortoise, which does not take any sustenance for
some time before it becomes torpid, and rejects
food also for a short period after it recovers
from its lethargic state. Martial says, of the
dormouse,—

> " Tota mihi dormitur hyems, et pinguior illo
> Tempore sum, quo me nil nisi somnus alit."

Now, though the dormouse does not become
fatter during torpidity, it may be considered as
in good condition, if not fat, before it commences its repose; and at its conclusion, it has
lost somewhat of its *en bon point*. It is certain
that the alpine hunters of Switzerland find the
marmots fat in their holes, and it is also certain
that after they leave them they are often seen

emaciated. Spallanzani is of opinion that dormice lose weight by torpidity, and Mr. Cornish states that bats and dormice lose from five to seven grains during the torpidity of a fortnight.

We may therefore infer safely, that digestion has ceased entirely, but that there still may be a slow expenditure or waste, perhaps cuticular, in the shape of insensible perspiration.

V. Animals in a torpid state seem insensible to external agencies. The dormouse may be tossed up as a ball, or thrown to a distance, without any change of state. Volta and Spallanzani could not rouse the marmot from its torpidity by the electric spark, and it was only temporarily disturbed by the shock from a Leyden jar. Wounds even have been inflicted, and their limbs broken, yet torpidity remained unhinged, as in bats, &c.

External stimuli, therefore, do not affect the torpidity of animals. Heat and air are the only agencies which rouse them from this death-like lethargy.

" The tribe of quadrupeds," says Dr. Reeve,

" have the habit of rolling themselves into the
form of a ball during ordinary sleep; and they
invariably assume the same attitude when in
the torpid state, so as to expose the least possible surface to the action of cold: the limbs are
all folded into the hollow made by the bend-
ing of the body: the clavicles and the sternum
are pressed against the fore-part of the neck,
so as to interrupt the flow of blood which sup-
plies the head, and to compress the trachea;
the abdominal viscera and the hinder limbs are
pushed against the diaphragm, so as to interrupt its motions, and to impede the flow of
blood through the large vessels which penetrate
it, and the longitudinal extension of the cavity
of the thorax is entirely obstructed." Such is
the phenomenon of the animal cradled in torpidity. Contrary to the opinion sustained by
Mr. Gough and others, animals prepare for
this singular state from instinctive feeling or
animal presentiment. "The stork knoweth her
appointed time." The silk-worm prepares its
cocoon, where, in an apparently lifeless image,
it undergoes its remarkable transformation.

The swallows,* " intelligent of seasons," con-
gregate in myriads at a specific period, and
wing their way to distant lands, where insects
still chequer the atmosphere and flicker in the
sun-beam, or eddy in the twilight, or dance on
the surface of the stream.—

> " Who bids the stork, Columbus-like, explore
> Heavens not its own, and worlds unknown before?
> Who calls the council, states the certain day,
> Who forms the phalanx, and who points the way?"—

I have been informed, by individuals who
have kept nightingales, that for a day or two
(even after they had been domiciled for years,)
before their usual and fixed period of migration,
they become restless, and beating themselves
violently against the bars of the cage, would
inevitably destroy themselves, but for the precaution taken of lining the cage with green
baize, or other soft material.   Just so it is with
torpid animals, they retire *before* their proper
food is withdrawn, and ere yet the severe season

---

* See Mr. Farster, " On the Brumal Retreat of the Swallow." London. Second Edition.

has arrived, prepare for its approach, by every precaution necessary to protect and preserve them. " Not one of them is forgotten" before Him, without whose permission " not a sparrow falls." This presentiment was necessary to fulfil the purposes and ends of their being, and therefore the peculiarity was superadded; and thus gifted by Providence,—

" They reason not contemptibly."

We rest secure in this entrenchment of opinion, and acknowledge here, as in other departments of natural history, a provident care which excites our wonder and our delight. A peculiar organization is no doubt prepared by creative wisdom for this singular display, and Sir Anthony Carlisle informs us, that animals of the class *mammalia*, which become torpid in winter, have at all times a power of subsisting under a confined respiration, which would destroy other animals not having this singular habit. He discovered a peculiar structure of the heart and its principal veins. We may safely draw from our premises the following

P

conclusions, and they certainly seem inductive and warranted inferences:—

I. Some animals as *naturally* pass by instinctive feeling into the state of torpidity, as others, such as the swallow, cuckoo, &c. migrate, till Providence in due time restores to them their natural and necessary food. It is not doubted that there is a peculiar organization fitted for these extraordinary functions, and subservient to this peculiar phenomenon. Night is the season for common and temporary repose—the winter of the year for this extraordinary sleep.

II. In torpidity the functions of respiration are completely suspended and at rest; but as some of the torpid animals would seem to pass into this state, fat, and emerge " lean and ill-favoured," it may be inferred that some slow and peculiar function, substitutary, is exerted. Whether this fat is absorbed into muscle, or evolved in another form, through the medium of some slow process, cannot be determined.

III. From the fact of torpid animals, as the hedge-hog, dormouse, &c. wrapping themselves up into a ball with *dried* leaves, hay, or other *non-conducting* substance, we may infer that moisture or damp, as well as a medium very low in temperature, would be injurious or fatal. An insulation, and a defence against these, is thus however provided. A sudden warmth, too, might, from its abrupt transition, cause the torpidity to merge in death: as an elevated temperature, when incautiously applied, destroys the frozen limb.

IV. The pre-disposing causes of torpidity are cold and want of food, or proper nourishment; and torpid animals instinctively provide against their advent. Abstraction of liquid may also minister to this phenomenon. To these may be added the absence of the stimulus of light and stagnation of air, or its slow currency and intercepted circulation.

V. Torpidity may be arrested or suspended by a proper temperature and generous diet, or nourishment, conjoined with free ventilation or currency of air. It is not improbable, however, that the character and habit of the animal may be thus changed, and perhaps the natural term of its life abridged.

The tortoise may be occasionally met with in gardens in this country. The *testudo geometrica* I have certainly seen here; but the occurrence is rare. One of three tortoises (the common), laid three eggs in a garden at Montrose—one of these I forwarded to Professor Jameson, of Edinburgh. The size to which this creature occasionally attains is quite monstrous. I remember, some years ago, to have seen one, then semi-torpid, exhibited near Exeter 'Change, London, which weighed, if I recollect aright, *several hundred weight:* 1: Its shell was proportionally thick, and its other dimensions bore a corresponding ratio. It was stated to be about 800 years old. In the library at Lambeth Palace is the shell of a land tortoise, brought there about the year 1623 : it lived until 1730, and was killed by the inclemency of the weather during a frost, in consequence of the carelessness of a labourer in the garden, who, for a trifling wager, dug it up from its winter retreat, and neglected to replace it. Another tortoise was placed in the garden of the Episcopal Palace at Fulham, by Bishop

Laud, when Bishop of that See, in 1628: this appears to have died a natural death in 1753. It is not known what were their several ages when placed in the gardens. That of which I am about to give an account, I saw in the Bishop's garden at Peterborough, adjoining the Cathedral, in the summer of 1813. It died only four or five years ago. *Why* this Episcopal predilection is a question perhaps not unworthy antiquarian research! The testudo græca is found in the Island of Sardinia—generally weighing four pounds, and its usually computed age is about sixty years.

From a document belonging to the archives of the Cathedral, called the *Bishop's Barn*, it is well-ascertained that the tortoise at Peterborough must have been about 220 years old. Bishop Marsh's predecessor in the See of Peterborough had remembered it above sixty years, and could recognize no visible change. He was the seventh Bishop who had worn the mitre during its sojourn there. If I mistake not, its sustenance and abode were provided for in this document. Its shell was perforated, in order to attach it to a tree, &c. to limit its ravages among the strawberry borders.

The animal had its antipathies and predilections. It would eat endive, green pease, and even the leek—while it positively rejected asparagus, parsley, and spinage. In the early part of the season, its favourite pabulum were the flowers of the dandelion. (*leontodon taraxacum*), of which it would devour *twenty* at a meal; and lettuce (*lactuca sativa,*) of the latter a good sized one at a time, but if placed between lettuce and the flowers of the dandelion, it would forsake the former for the latter. It was also partial to the pulp of an orange, which it sucked greedily.

About the latter end of June, (discerning the times and the seasons,) it looked out for fruit, when its former choice was forsaken. It ate currants, raspberries, pears, apples, peaches, nectarines, &c., the riper the better, but would not taste cherries. Of fruits, however, the strawberry and gooseberry were the most esteemed: it made great havoc among the strawberry borders, and would take a pint of goose-

berries at intervals. The gardener told me it
knew him well—the hand that generally fed it—
and would watch him attentively at the gooseberry bush, where it was sure to take its station
while he plucked the fruit.

I could not get it to take the root of the
dandelion, nor indeed any root I offered it—as
that of the carrot, turnip, &c. All animal food
was discarded, nor would it take any liquid;
at least neither milk nor water; and when a
leaf was moist, it would shake it to expel the
adhering wet.

This animal moved with apparent ease,
though pressed by a weight of 18 stones; itself
weighed 13½ lbs. In cloudy weather, it would
scoop out a cavity, generally in a southern
exposure, where it reposed, torpid and inactive,
until the genial influence of the sun roused it
from its slumber. When in this state the eyes
were closed, and the head and neck a little
contracted, though not drawn within the shell.
Its sense of smelling was so acute, that it was
roused from its lethargy if any person ap-
proached even at a distance of twelve feet.

About the beginning of October, (or latter
end of September,) it began to immure itself,
and had for that purpose for many years selected a particular angle of the garden; it
entered in an inclined plane, excavating the
earth in the manner of the mole; the depth to
which it penetrated varied with the character of
the approaching season, being from one to two
feet, according as the winter was mild or severe.
It may be added, that for nearly a month prior
to this entry into its dormitory, it refused all
sustenance whatever.   The animal emerged
about the end of April, and remained for at
least a fortnight before it ventured on taking
any species of food.   Its skin was not perceptibly cold:* its respiration, entirely effected
through the nostrils, was languid.   I visited
the animal, for the last time, on the 9th June,
1813, during a thunder-storm: it then lay

---

* Dr. Davy took the temperature of the *testudo geometrica*
at Cape Town, in May—air 61°; the animal 62°.5.   At
Columbo, in Ceylon, on 3d March, the temperature of a
larger specimen was 87°, while the air was 80°.

under the shelter of a cauliflower, and apparently torpid.

It is very singular that the lettuce and dandelion should find such predilections with the tortoise. The lactescent juice of the former, from the opium it contains, is powerfully narcotic, and I have found that the extract. taraxici, applied to the sciatic nerves of a frog, acted in a similar manner to opium, by suspending voltaic excitement. It is also remarkable that these should have been rejected when the fruit season commenced, and the strawberry and gooseberry take precedence. Its antipathy to cherries is equally curious, and not less so its aversion to fluids; in which last, however, we have an analogy in the *alpaca*, &c.

On the whole, that narcotics or sedatives should take precedence, of all other kinds of food, in the former part of the season, and those that act a different part, in the animal economy, toward the autumn, is certainly surprising.

As a proper sequel to the preceding, I may add my remarks on

The Temperature of the Skin of the Dormouse.

In the beginning of 1824, I received two dormice from a friend in Derbyshire, and commenced a series of experiments on the temperature developed by the skin. One of these I accidentally lost, it having escaped from confinement; and I was shortly necessitated, from various avocations, to resign the prosecution of my researches with the other. The following is a note of the temperatures as recorded:—

31st January, 1824, Chesterfield, Derbyshire, at 25 minutes past 7, P. M.; air of the room, 48° Fahrenheit; temperature of the dormice under the breast, 103° Fahrenheit. I soon after lost one of my prisoners.

At Hull, Yorkshire, 14th February, at half-past 8, P. M.; air 51° Fahrenheit; temperature under the breast, 62°.5 Fahrenheit. *The animal semi-torpid.*

		Air.	Under breast.
Feb. 15, At 1 h. 15 min. P. M.	46°	104°	
... ... 8 h. 30 min. P. M.	47°.5	69° Semi-torpid.	
... ... 3 h. 30 min. P. M.	52°	102°.5	
... 19, ... 2 h. P. M.	56°	99°	
... 21, ...10 h. 30 min. P. M.	54°.5	102°	
... 22, ...12 h. 30 min. P. M.	57°	97°	

On the 14th and 15th of February, the dormouse was roused from its apparent death by heat, cautiously applied.

The box which contained the dormice had a partition. One compartment contained fresh moss, well dried, in which the animals reposed *during day,* having formed for themselves a somewhat elliptical nidus. Two openings, with slides, conducted into the *outer court,* where the dormice had their food prepared for them, consisting of wheaten bread, (sometimes softened with water,) and a basin of milk. Great attention and care were bestowed on them, and the food daily supplied. The sliding pannels were shut when the compartments were cleaned, it being easy to expel them from the one to the other, and thus prevent their escape.

Though their cage was frequently in darkness during the day, the *night season* was the exclusive period in which they took food. One of them had a singular expedient when the liquid was too low in the basin. It dipped its brushy tail (somewhat resembling that of a fox,) into the dish, and carried the milk in this manner to its mouth. When the dormice are torpid, they may be tossed up into the air like a ball, (they are rolled up like one,) without discovering any index of motion or change. By keeping the dormouse in a proper temperature during the winter, its brumal torpidity may be entirely prevented; but in this case it will not outlive the following year. The dormouse is fat and in good condition when it enters into torpidity, and it issues from this state miserably lean. My dormice were extremely timid, yet they may be so tamed as to run about the table and lick the hand that feeds them. As to their sense of hearing, I found them peculiarly affected by the higher notes, or shrill tones. The eyes were like those of albinos, and injured by strong light and exposure to day.

Dr. Davy gives us the following details of temperature, which approach that of the dormouse as stated :—

			Air.	Animal.
Columbo, Ceylon,	19th Oct.	Squirrel,	...81°...	102°
	8th Feb.	Common Rat,	...80°...	102°
	16th June.	Common Hare,	...80°...	100°
	4th Nov.	Ichneumon,	...81°...	103°
	26th Feb.	Jungle Cat,	...80°...	99°

Dr. Davy, I think, justly infers that the temperature of the human species *increases* in passing from a cold, or even a temperate climate, into one that is warm; and I think, too, that I am warranted, from my own immediate observations and experiments on myself, to add, that the temperature of man *rises* with *increase of elevation.* On the summit of the Simplon, with the ball of the instrument below the tongue, the temperature was 100°.5; on Mount Cenis, 101°; and, on the Great St. Bernard, 102° nearly. The temperature of animals will, no doubt, be modified by, or have some determinate relation to peculiarities in physiological character.

THE END.

JAMES CURLL, PRINTER, GLASGOW.